Theoretical Biology

Epigenetic and Evolutionary Order
from Complex Systems

Theoretical Biology

Epigenetic and Evolutionary Order
from Complex Systems

Brian Goodwin
and
Peter Saunders

Editors

EDINBURGH UNIVERSITY PRESS

©Edinburgh University Press 1989
22 George Square, Edinburgh
Set in Lasercomp Plantin
at The Alden Press, London
and printed in Great Britain by
Redwood Press Ltd, Trowbridge
British Library Cataloguing
 in Publication Data
Theoretical biology: epigenetic
 and evolutionary order.
1. Biology
1. Goodwin, Brian II. Saunders, Peter
574
ISBN 0 85224 600 5

Contents

Preface

In recent years there has been a growing feeling that it is time to create a forum for discussion of the issues that were raised and explored in the series of meetings organized by C.H. Waddington in the years 1968–71, from which emerged the four influential volumes *Towards a Theoretical Biology*. For Waddington they continued the tradition of discussion which he, along with others including Joseph Needham, J.H. Woodger, and J.D. Bernal had started when they established the Theoretical Biology Club, which met in Cambridge in the 1930's. Waddington and Needham came very close to establishing in Cambridge during this period an institute for interdisciplinary studies in biology under the patronage of the Rockefeller Foundation. Their objective was to bring together biologists, physicists, chemists and mathematicians to work on the basic problems of evolution, genetics, development, and molecular biology. In the end they did not succeed, though for reasons unrelated to the merits of the proposal or the scientific credentials of those involved, and to this day no such institute exists.

There still seems little prospect of setting up the sort of centre that Waddington and his colleagues envisaged. On the other hand, there are now scientists in many countries who are actively engaged in the programme they had in mind. Recognising this, a group of biologists and physicists at the Universidad Nacional Autonoma de Mexico, who had already formed an interdisciplinary group to discuss some of the fundamental questions in biology, decided that the time was ripe to continue the Waddington programme in an international setting. They saw that this could be done by creating a network of scientists who could interact and gather together for meetings and particular tasks, even if no permanent centre could be provided.

As a first step, Germinal Cocho contacted Brian Goodwin, who had long-standing Mexican contacts since a sabbatical visit to UNAM in the early '70's, and a meeting was arranged. Invitations were sent to an international group of biologists, physicists and mathematicians to gather in Oaxtepec, Mexico, in September 1987. Since the main point of such a meeting is discussion, the total number of participants was restricted to about thirty-five, and the programme was left flexible. Continuity with the Waddington meetings was achieved not only in spirit: Brian Goodwin, René Thom, Lewis Wolpert and Christopher Zeeman who had been participants at one or more of the Serbelloni meetings were also at Oaxtepec; and Stuart Kauffman, who was unable to come at the last minute, contributed a paper.

The conference also witnessed direct continuity of mathematical inspiration and the rare sight of mathematics in progress: Zeeman presented a theorem that he had just proved giving a new definition of stability for dynamical systems in n-dimensions, based upon but generalizing Thom's original notion of structural stability for gradient systems. He added the conjecture that this is a generic property of dynamical systems, a conjecture for which López de Medrano was able to suggest a proof during the course of the meeting. This led to a classification of generic differential equations and their bifurcations.

The theme of generic properties in biology could well be singled out as the dominant strand of the whole conference. This style of thought, the very essence of mathematics and physics, is still regarded with deep suspicion by many biologists, who see it as a superfluous, logical framework artificially imposed on what is really an empirical subject. Many of the papers in the volume address this question in one form or another. The deep underlying issue is intelligibility: if a subject has no generic properties, no generative principles that apply throughout and give a unifying logical structure to diverse empirical phenomena, then the subject matter cannot be grasped rationally. This is one of the tensions between theoretical and empirical biology. It also underlies the persistent dialectic between evolution, which is generally perceived to be a random process, and development, which many, like Waddington, see as having an intelligible structure. Much hangs on the resolution of these issues, which is why they have such prominence in this volume.

Despite, or perhaps because of, its flexible structure, the meeting was stimulating and informative, with many fascinating discussions and keen debates. We have, however, decided not to try to keep as closely to the proceedings of the meeting as Waddington did in the series he edited. Instead, while the papers that make up this volume are for the most part based on talks that were given at Oaxtepec, they have been written in a more formal and complete style, and have benefited from discussion both at the meeting and subsequently. We have resisted the temptation to try to produce a *Towards a Theoretical Biology 5*; it is a measure of the success of the original series that something different now seems more appropriate.

Ambience and atmosphere add significantly to the quality of intellectual exchange. The Villa Serbelloni on Lake Como was an idyllic setting for the Waddington meetings. It was matched by the exotically beautiful Mexican conference site of Oaxtepec, with Popacatepetl visible in the distance. The legendary hospitality and generosity of our Mexican hosts also played an important part in the success of the meeting. It may be difficult to equal these locations in future, but it is intended that discussion in this style and on these and related topics should continue, involving an international network of interested individuals with a variety of professional affiliations, gathering at fairly regular intervals. The appropriate organisation will then develop and evolve, and we shall understand our subject by experiencing it. This volume is offered as a link in a continuing tradition, and with a particular dedication to C.H. Waddington, to whom so many of us are indebted for his imagination and breadth of vision.

B Goodwin
P Saunders

Contributors

F. Alonso-deFlorida
Departamento de Biofisica y Biomate-
maticas
Laboratorio de Biofisica Experimental,
Instituto de Investigaciones Biomedi-
cas, U.N.A.M.
Apartado Postal 70990, Cd. Univer-
sitaria C.P.
04510 Mexico
D.F. Mexico

Dr Germinal Cocho
Instituto de Fisica
Universidad Nacional Autonoma de
Mexico
Apdo. Postal 20-364
Delegacion Alvaro Obregon
01000 Mexico
D.F. Mexico

J. Collado Vides
Instituto de Fisiologia Celular
Universidad Nacional Autonoma de
Mexico
Apdo. Postal 565-A
Cuernavaca, Mor
Mexico

A. Coutinho
Unité de Immunobiologie
Institut Pasteur
23 rue du Dr. Roux
75015
Paris, France

Professor P.C.W. Davies
Dept of Physics
University of Newcastle
Newcastle-upon-Tyne
England

Dr J. Chris Eilbeck
Dept of Mathematics
Heriot-Watt University
Riccarton
Edinburgh EH14 4AS
Scotland

Professor Brian C. Goodwin
Dept of Biology
The Open University
Walton Hall
Milton Keynes
MK7 6AA
England

Dr Mae-Wan Ho
Biology Dept
The Open University
Walton Hall
Milton Keynes
MK7 6AA
England

Dr Bernardo Huberman
Xerox Corporation
Palo Alto Research Center
3333 Coyote Hill Road
Palo Alto
Ca 94394
USA

Professor Stuart A. Kauffman
Dept of Biochemistry and Biophysics
School of Medicine G3
University of Pennsylvania
Philadelphia 19174
USA

Carmela Kubal
Dept of Mathematics
King's College
Strand
London WC2R 2LS
England

Dr Francisco Lara-Ochoa
Centro de Investigacion Sobre
Fijacion de Nitrogeno
Universidad Nacional Autonoma de
Mexico
Apdo Postal 565-A
Cuernavaca, Mor
Mexico

Miguel A. Jimenéz-Montaño
Centro de Investigaciones Biologicas
Universidad Veracruzana
Apartado Postal 302
Xalapa Ver.
Mexico

A.A. Minzoni
Departamento de Matematicas y
Mecanica
I.I.M.A.S.
U.N.A.M.
C.P. 01000 Mexico
D.F. Mexico

Dr Jose L. Rius
Biologia Teorica Y Fisicoquimica
Biologia
Crupo Universitario Interdiscipilinario
Apartado Postal 565-A
Cuernavaca
Mexico

Vincente Sánchez-Leighton
Hyphen Informatique
7 rue Leopold Bellan
75002 Paris
France

Professor Peter T. Saunders
Dept of Mathematics
King's College
Strand
London WC2R 2LS
England

Professor Michael A. Savageau
Dept of Microbiology and Immunology
6643 Medical Science Building II
Ann Arbor
Michigan 48109
USA

Professor René Thom
Institut Des Hautes Etudes Scientifiques
35 Route de Chartres
91440 Bures-sur-Yvette
France

Professor Lynn E.H. Trainor
University of Toronto
Dept of Physics and Medicine
Toronto
Ontario
Canada M5S 1A7

Professor Francisco J. Varela
Ecole Polytechnique
C.R.E.A.
1 rue Descartes
F-75005
Paris
France

Professor E. Christopher Zeeman
Mathematics Institute
University of Warwick
Coventry
Warwickshire
CV4 7AL
England

Introduction

BRIAN C. GOODWIN and PETER T. SAUNDERS

Because theoretical biology was still in its early stages as a discipline, the topics discussed in the Serbelloni meetings and the four volumes that arose out of them covered a very wide range of subjects and approaches. Underlying all this, however, can be seen an attempt at integration, which was also characteristic of much of Waddington's own work. One of his major preoccupations was to achieve a synthesis of development and evolution, to resolve what he experienced as a conflict between the ordered transformations of epigenesis on the one hand and the randomness of neo-Darwinism on the other. He saw the basis for this in terms of the potential of developmental processes for adaptive response to environmental influences and the stabilization, or as he preferred to call it, canalization, of particular pathways through the epigenetic landscape to the adult form. For him this idea was not a solution but a guide to asking the right questions. It defines a research programme, one which involves a search for the principles of self-organization that underlie devlopmental processes and their relation to the functional requirements of organisms in specific environments.

The theme that unites the contributions to this volume is just this problem of understanding the relationships between epigenetic and evolutionary order. Readers who are not biologists may find it strange that anyone should have to direct attention to such an obviously important question, but such has been the predominance of the neo-Darwinist view in recent years that the study of biological order as a thing in itself rather than an epiphenomenon arising out of selection has been largely neglected. Biologists have thus fallen behind their colleagues in other disciplines, and it is significant that almost all the contributions to this volume draw on work that has been done in other fields, in which order has remained a central preoccupation.

MATHEMATICS

Of all disciplines, the one that deals most directly with order is mathematics, and so the volume begins with four mathematical papers. In the first, the impact of Waddington's vivid geometrical metaphor of the epigenetic landscape on René Thom's profound insights into the singularities of smooth maps and thus the classification of gradient and other dynamics is described by Thom himself. This amply demonstrates the strength of the research programme that Waddington set out, for the success of any such programme is to be measured by the importance of the work it stimulates, especially in directions which may well not have been envisaged at the outset. As Thom has acknowledged elsewhere, in this respect mathematics so far owes more to biology than conversely.

The continuing development of the work stimulated by Waddington and initiated by Thom is described in the second paper. Thom originally hoped to be able to classify all dynamical systems, but while catastrophe theory shows the way forward and accomplishes a great deal, it has since been shown that the concept of structural stability, on which it depends, is too restrictive. There are, loosely speaking, too many systems that are not structurally stable. Zeeman therefore proposes a new basis for classification which extends Thom's work in a potentially very fruitful way. Next, to show that biology can expect some return from the important stimulus it has given to mathematics, Saunders and Kubal use catastrophe theory to suggest typical properties of developmental systems beyond what can be revealed by the useful but necessarily over-simplified picture of the epigenetic landscape. It is fitting that they derive a testable prediction about phenocopies, because the study of phenocopies had a very strong influence on Waddington, especially in the development of his theory of genetic assimilation.

The other major influence on the mathematical modelling of epigenesis has been A.M. Turing. His life-long interest in biology, and especially in biological form, inspired him to the discovery (Turing, 1952) that a simple combination of biochemical reactions and diffusion could generate patterns in an initially homogeneous domain. The result may seem counter-intuitive, since diffusion usually brings about smoothing, but it arises out of bifurcations that break the symmetry and generate order. Eilbeck's studies of reaction-diffusion systems show the rich and robust sequences of bifurcations and consequently the large degree of self-organization that can occur even in quite simple systems.

ORGANIZATION AND SELF-ORGANIZATION

The relationship between epigenesis and evolution is explored in the papers by Savageau, Kauffman and Goodwin. Savageau points out that gene regulatory circuits can be divided into two categories according to whether the control is positive or negative. He shows that the type of control which is stable with respect to mutational perturbation depends on the environmental pattern of nutrient availability, and that in any real situation it is the stable one that is found. His work demonstrates how explanations of evolutionary phenomena depend on understanding the set of possible epigenetic states and an analysis of the relative stability of these states in particular environments.

Kauffman examines the influence of natural selection on a class of complex systems that simulate the very rich internal interactions of biological systems. He finds that the order in such self-organizing systems prevents natural selection from moving them far from their generic states. Thus selection can only choose from among the alternatives placed before it by the dynamics; it is not a creative force. Goodwin examines natural selection in connection with the evolution of epigenetic forms such as the gastrula. He argues that not only are historical selective scenarios not susceptible to experimental test, they are inherently incapable of accounting for origins. An adequate explanation requires an understanding of the generative principles based on an experimental study of the processes of development, the identification of the set of alternative forms that can be generated, and finally stability analysis of these alternatives. Natural selection deals with only the last of these.

The idea of self-organization is carried a step further by Davies, who suggests that highly complex non-linear systems may exhibit regularities that cannot be deduced from the known laws of physics but would constitute a new level of fundamental order. What we perceive as the progressive local increase in order as the cosmos develops from the big bang could be the result of a new principle of organization that applies to the inanimate as well as to the animate. The papers of Varela, Sanchez-Leighton and Coutinho and of Huberman explore the implications of self-organization in relation to order and adaptability in particular classes of complex networks. The former is based on studies of the network properties of the immune system and its capacity for "learning" appropriate responses. The latter involves an investigation of adaptability in complex systems with different types of organization, both distributed and hierarchical. Both papers break significant new ground in describing the generic properties of complex networks in relation to learning and to adaptive behaviour.

PHYSICS

There are essentially two ways in which theoretical biology draws on its more advanced neighbour, theoretical physics. Biologists can use actual results from physics, or they can try to carry over ideas and concepts and use them, suitably modified, in their own work. Examples of both are included in this volume.

Trainor investigates how the field concept, so fundamental and successful in physics, can be applied to certain problems in morphogenesis, especially to the study of limb regeneration in amphibians. He shows how a vector field description permits a more successful analysis of the complicated observations of supernumerary limb production than do more conventional approaches. Another topic which has been well developed in physics is the study of phase transitions, and Lara Ochoa applies these ideas to biological systems, in particular to membrane behaviour modified by ionic influence.

A characteristic feature of any organized sytem is that it possesses a means by which information and energy can be transferred quickly and effeciently among the components. Ho discusses the role that coherent excitations might play in this, not only in such processes as muscle contraction but also during development, i.e. during the actual creation of the organization. She also reviews the growing evidence for the phenomenon in biology, including the recently recognized dangers associated with low intensity, low frequency electromagnetic radiation.

Finally, Cocho and Rius apply the cellular automata model to the study of pattern formation. They demonstrate how simple local interaction rules can generate complex patterns that closely resemble the actual pigment distribution patterns in a variety of vertebrate species. Such models share with field models the property that global spatial order can arise out of local interaction.

LINGUISTICS

The last two papers in this volume use concepts developed in linguistics, a subject which lies outside the natural sciences but in which the study of order has been carried to a very sophisticated level. Jiminéz-Montaño points out that an important problem of molecular biology is to understand the structure of the language of the genetic code, i.e. to find the rules that determine which sequences of symbols ("sentences") are meaningful and which are not. He then shows how ideas from structural linguistics can contribute towards an understanding of how the information contained in DNA, RNA and proteins can code for specific three dimensional structures. Collado Vides, who stresses that it is the relations among the nucleotides rather than the individual nucleotides themselves that are important, draws on the concept of a generative grammar in his study of the problem of genome regulation and regulation of gene expression.

These papers form an appropriate conclusion to the volume, because the *Towards a Theoretical Biology* series also ended with a discussion on the relevance of linguistics to biology. In the very last paragraph Waddington (1972) wrote, 'To a biologist, therefore, a language is a set of symbols, organized by some sort of generative grammar. . . And it is language in this sense that I suggest may become a paradigm for the theory of General Biology.'

Twenty years after the Serbelloni meetings, theoretical biology has still not been unified, and we cannot yet speak of a paradigm for the subject in the way that Waddington hoped for. Perhaps there never will be a single paradigm; perhaps the subject matter is too complex and varied for that. That is not to say, however, that theoretical biology has not matured as a discipline and become somewhat more like what Kuhn (1962) calls a normal science. The contributions to this volume still represent a number of different approaches, certainly more than one would expect to find in a collection of papers in theoretical physics. All the same, the reader will find more unity here than in any of the volumes that Waddington edited. The integration that he strove towards may never be achieved, and the advances may not happen quite as he expected, but there has been progress and much of it has been in the directions that he foresaw. Order and organization are once again primary issues in biology, and epigenetics is being recognized as of great importance for an understanding of evolution. That, more than the material in this volume, is Waddington's real memorial.

REFERENCES

Kuhn, T.S. (1962). *The Structure of Scientific Revolutions*. Chicago, University of Chicago Press.
Turing, A.M. (1952). The chemical basis of morphogenesis. *Phil. Trans. Roy. Soc.* B237, 37–72.
Waddington, C.H. (1972) (ed.). *Towards a Theoretical Biology: 4 Essays*. Edinburgh, Edinburgh University Press.

1. An Inventory of Waddingtonian Concepts

R. THOM

A little more than twenty years after his death, it may not be too early to assess the impact of C.H. Waddington's work and ideas on the later developments of biology. I shall leave aside his purely experimental work. Quite obviously, his long activity as an embryologist led him to many interesting results: let me quote his 1937–40 work on chick embryology, where he tested the inducing effect of the primitive streak on different locations of the graft, and his later work on *Drosophila*, dealing with 'genetic assimilation'. Whatever merit his laboratory achievements may be credited with, I suppose almost anybody would agree that his importance in the historical development of twentieth-century biology lies elsewhere. It is founded in his great ability to develop concepts, and to coin for them appropriate names. Here is a list (perhaps incomplete) of words which he proposed as neologisms:

Canalization
Creods
Epigenetics–Epigenotype
Epigenetic Landscape
Evocation–Individuation
Genetic Assimilation
Homeorhesis

One may wonder why I apparently attach so much importance to words, and not to ideas proper. There is a simple reason for this: It is always extremely difficult to assess the originality of an author as he is proposing some novel idea. My own experience has taught me that one can never be considered as the 'owner' of an idea. Ideas are always – to some extent – created before you find them, and you will discover them, under a somewhat different form, in the published work of people you have no reason to believe had any knowledge of your own publications. But you may be the 'owner' of a word which you created, and which will follow you throughout your life (and hopefully will outlive you). These considerations are not here to belittle C.H. Waddington's originality – which was, in fact, very great among his biological colleagues. But this originality was relatively slow to develop. Although in his youth he was deeply impressed by A.N. Whitehead's philosophy, I think it is fair to say that his basic conceptions in biology were those of an orthodox Darwinian, at least until around 1960. His conceptions about genes, and the interaction between genes and cytoplasm were of a relatively classical cybernetical nature. I suspect that the rapid development of molecular biology in the decade 1960–70 was the triggering factor which made him develop his 'marginalistic' tendencies in a more systematic way. But it is not my purpose here to describe 'the evolution of an evolutionist'. Instead, let us consider the fate of all these words in subsequent years.

EVOCATION–INDIVIDUATION

'Individuation' was the first term proposed by Waddington to describe the overall organising effect of an 'organizer' (such as the celebrated dorsal lip of the blastophore in amphibians). I do not think this attempted 'word launching' was a very successful one. It is true that in 'normal' embryological development, some tissues have the power to trigger in nearby tissues the formation of organs morphologically (and frequently functionally) associated with the organ normally developing from this centre. It is true that this process has frequently some 'individual' character, but experiments have shown that induction in a competent tissue may be started experimentally by the most heteroclite material. The host of the graft also has some power to influence the induced morphologies. As Waddington pointed out to me (I suppose it was at one of the Serbelloni meetings) the distinction between inducer and induced tissue is somewhat arbitrary, as frequently the induced tissue will direct the original inducer in a direction compatible with its own 'competence'. What is in question is the problem of 'individuality', a very tricky problem, taken 'metaphysically' or 'scientifically'. Hence I do not think that, at the time, Waddington himself was very interested in the future survival of his 'individuation'.

'Evocation' was, of course, a much better word, as we have to describe this incredible variety of possible inducers in the neurogenic induction of vertebrates. As such, I do not know whether it is still much in use. Of course, the concepts of 'competence' and of 'loss of competence' are obviously more precise and useful. It should be pointed out here that neurogenic induction is perhaps the only one to show this amazing variety of 'evocators'. Later inductions, such as those of limb buds and of organs such as the kidney, lung, or ears, require much more specific 'evocations'.

EPIGENETICS–EPIGENOTYPE

These words were coined by Waddington in his 1940 book *Organizers and Genes*. 'Epigenetics' was the result of a marriage synthesis between 'Epigenesis' (the classical older theory opposed to 'preformation' in embryological theory) and 'genetics' (the same applies to 'epigenotype'). 'Epigenetics' had a relatively fortunate destiny. Whenever anybody said that embryological development was 'coded in the genes', a few experimentalists were dissatisfied with this somewhat dogmatic statement, and were inclined to emphasise the importance of local morphogenetic factors such as mechanical strains on tissues (following the Entwicklungsmechanik of Roux), the contact with nearby tissues, local environmental influences like external gradients, etc.; hence the need for a word subsuming all these local events. Personally I am inclined to think that the distinction in embryology between 'genetic' and 'epigenetic' events is a moot question (perhaps as ill-defined as the classical problem of distinguishing inherited from acquired characters). I suspect that the only pertinent question to raise has to do with causal mechanisms. If you were to follow Aristotle's theory of causality (four types of causes: material, efficient, formal, final) you would say that from the point of view of material causality in embryology, everything is genetic – as any protein is synthetised from reading a genomic molecular pattern. From the point of view of efficient causality, everything is also 'epigenetic', as even the local triggering of a gene's activity requires – in general – an extra-genomal factor.

CANALIZATION, EPIGENETIC LANDSCAPE, CREOD, HOMEORHESIS

These may be considered the most important Waddingtonian concepts. *Canalization* was the great discovery already presented in Waddington's (1940) book, *Organizers and Genes*. It describes any kind of process whose temporal evolution is buffered against external perturbations. There is apparently no strict mathematical equivalent to this concept, as the classical 'structural stability' is of global topological nature, whereas 'canalization' has a metric and local character. The concept had apparently a large local success (at the celebration of Waddington's fiftieth anniversary in 1955 at the Edinburgh Institute of Animal Genetics), it was a 'magic word', the theme of a song). Among biologists, however, it was considered as a useful concept but with no explicit application.

The same observation applies also to the allied concept of *creod*. This word – from the Green $\chi\rho\acute{\eta}$: (it is necessary), and $\acute{o}\delta o\varsigma$ (the way) – was meant to describe any kind of obligatory path for a process of an arbitrary nature. As such, although in principle the concept was quite pertinent, it turned out to be a failure. When I wrote *Structural Stability and Morphogenesis* (1972), I adopted the word, but in my mind for a fairly precise situation. The concept as Waddington proposed it was obviously too general, as nothing was said about the nature of the process and how it could be described. For instance, it was not clear whether the process involved continuous changes only or, on the contrary, morphogenetic events due to the apparition of phenomenological discontinuities. Because of this too general, fuzzy setting, the *creod* concept had little relevance for verbally described processes. For these it is in general useless, as any concept involves in itself some internal regulation mechanism limiting its extension, and hence has some internal 'creodic' structure. When you drive on a motorway, you do not feel the need to think that you are following a creod. It seems to me now that situations of the 'creodic' type are to be subsumed into two major types; which I shall now describe.

CANALIZATION AND MORPHOGENETIC FIELD

As I see it, these are the two major situations involved. A morphogenetic field always implies the appearance of new discontinuities in the developing embryo (the creation or destruction of an 'anhomeomere' in Aristotelian terminology, or the changing topological qualitative type of an 'homeomere'). A process involving only some discontinuity in the growth without new cellular differentiation (e.g. a jump of the exponents in an allometric growth process) should not be termed a morphogenetic field. But it may be 'canalized', when seen as a kinetic process.

I hope I am not speaking only with the professional prejudice of a mathematician. I claim that the general theory of morphogenesis in biology has been plagued by vague conceptualizing and, accordingly, by a proliferation of ill-defined terminology. The notion of a morphogenetic field, which started at the end of the twentieth century with the work of Driesch, Child, Boveri, etc. fell into disrepute; first because of its mystical finalistic connotations (Driesch's entelechy), but also, I believe, because of its inherent imprecision. Had people realised that the notion of a morphogenetic field is a purely descriptive notion, by itself devoid of any explanatory power, then a lot of trouble could have been avoided. One has to thank Lewis Wolpert for having, with his 'positional information' made acceptable to biologists a notion which any mathematician would call a co-ordinate system, or a map. Add to that the notion of a model polyhedral set K in the target Euclidean space defining the support of appearing discontinuities (the catastrophe set in catastrophe theory, the anhomeomeric parts in Aristotle's terminology (*De Partibus Animalium*), and you have the general notion of a morphogenetic field). Let me say also that describing the anatomical organization of *Metazoa* should not require considering too fine detail. The cellular structure in general is too fine to play a major role in ontogenetic process (at least for large animals); in particular the fractal structures may be eliminated – or 'formally truncated' – as true fractals never exist in nature. Let us consider now the next Waddington notion.

THE EPIGENETIC LANDSCAPE

This was a special case of 'creod' when applied to embryological development. It emphasized the essentially diverging character of tissue differentiation and incorporated the 'canalising' character of the process in the metaphor of a tree of valleys running downhill. There was a certain ambiguity in the model's description; in some cases Waddington defined the 'width' of the landscape as a spatial co-ordinate, and in other cases he interpreted it as an abstract, functional parameter. These two interpretations may be unified if it can be shown that a given spatial co-ordinate in the embryo is 'a morphogenetic gradient' having an overall functional meaning. In the picture of the epigenetic landscape it is interesting to reverse the height function by putting $V = -z$. Then the picture becomes a potential well, with the gradient trajectories of V defining valleys ending in the basin 0. Such a gradient dynamics was the object of a classical theory in mathematics (the so-called Morse theory). Moreover, the inner geometry of the valleys descending into the sink defines a tree structure, reminiscent of well-known graphs of the generative grammar theory in linguistics. By considering the case where the V function has a degenerate minimum at the sink 0, I was immediately led (around 1965) to the notion of unfolding of a singularity, and hence to the basic theorems underlying catastrophe theory. There is no doubt that the model of the epigenetic landscape was for me the decisive clue to the discovery of catastrophe-theory models. I was also struck, in those years (around 1964) by Max Delbrück's theory of cellular differentiation in embryological development, seen as a choice between several stationary mitotic regimes of a cell: a model very similar in spirit to the epigenetic landscape, which Waddington apparently did not appreciate very much.

Is the epigenetic landscape fixed once and for all, or is it flexible under external stress? This problem led Waddington to his well known experiments on 'genetic assimilation'. A somatic variant of *Drosophila* (similar to the bithorax mutant phenotype) was initiated by thermal shocks on eggs of wild-type *Drosophila*, and then subjected to artificial selection. After some time a population of mutant *Drosophila* was created, which reproduced the bithorax phenotype without thermal shocks. This could have led to the idea that epigenetic phenocopies phenotypically equivalent to the effect of genetic mutation can be described as 'local deformations of valleys' in the epigenetic landscape. Of course, a standard Darwinian explanation may be given for the result of this experiment. (How else could it be, as quite likely Darwinism can never be wrong!) Waddington, in 1955–65, wanted to bridge his epigenetics with classical genetics. This led him to the well-known image of an improved epigenetic landscape, where genes were figured by poles, and their effects by tracting ropes attached to these poles. A very suggestive figure, which – I was told by specialists – never found an application. Recently, I heard classical biologists saying that the epigenetic landscape model did not have any usefulness in developmental theory. Such an opinion, I think, expresses the still universally widespread dogma that any result in embryology which does not involve the activity of a gene in an explicit manner is worthless.

HOMEORHESIS

The vocable, in itself, is not very much in use. It expresses the following situation: Let X be a flow (a vector field) in a space M with m_t the corresponding time evolution; a trajectory γ generated by a point μ_t is 'homeorhetic' if for a large neighborhood U of μ_{t_0}, any point $m \in U$ remains for increasing t, in a vicinity of the trajectory γ, and tends towards γ, as $t \to \infty$.

In mathematics, one says that two points $m_{t_0} m'_{t_0}$ are asymptotic for X if the distance $|m_t m'_t|$ tends to zero as $t = \infty$. It is known that for a *generic* (X), the set of points asymptotic to a given m_0 is in general a manifold $S_k(m_0)$ (the so-called *stable manifold* of m_0), and that these S_k form a foliated structure in M. The 'isochrons' defined by Winfree in his celebrated 'Killing the clock' experiment are stable manifolds in that sense: it is a partial 'homeorhesis'. Such phenomena of 'homeorhesis' are always found in the basins of attractors.

Perhaps the true definition of homeorhesis should be weaker. One shrinks the trajectory γ to a fixed time interval $(\tau \tau')$, and one requires that for any $m \in U$, the trajectory (m_t) approximates γ during this interval only. (It is a kind of 'narrows' for the flow). Phenomena of this kind are frequent in models of constrained dynamics, involving fast and slow dynamics. They can be found also by computer calculation on very simple differential systems (the 'rivers' phenomena described by M. and F. Diener). The underlying conceptualization is, however, still missing, and I feel that the homeorhesis concept still has much to offer to mathematicians.

CONCLUSION

There is no doubt that the ideas and the model proposed by C.H. Waddington in his 1940–50 books had a great impact on some minds – especially mathematically minded ones. On biologists proper, their influence was perhaps not very strong. After the blooming of molecular biology, most embryologists turned to biochemistry or molecular biology. Now a few of them return to classical embryology under the heading of 'genetics of development'. Waddington himself knew quite a lot of mathematics, but of a more classical nature; for more modern notions, he had to appeal to his intuition – which was great. But he did not know enough of topology and 'modern' differential analysis to appreciate 'sophisticated' models such as those of catastrophe theory. I think he was perhaps a bit sceptical about the possibilities of the mathematics he knew in biological problems (perhaps because of his inconclusive experience with the 'mathematical genetics' of the Fischer-Haldane school). His Whiteheadian philosophy, of course, led him away from reductionism; but he was perhaps too attached to the ambient empirical Anglo-Saxon tradition to accept Platonic entities in biology and to treat them 'Platonically' (i.e. 'mathematically'). I feel sure he would have been happy to see the present revival of experimental embryology and the corresponding resurrection of the field notion – an already striking phenomenon in recent years.

ACKNOWLEDGEMENT

The author wishes to acknowledge that he drew very useful information on Waddington's life and career from the 1987 Ph.D. thesis by Marie Thérèse Ponsot entitled C.H. Waddington ou 'L'évolution d'un évolutionniste', thèse présentée à l'Université de Paris I – Sorbonne, 1987. He wishes also to thank Dr K. Millett, who reread the manuscript in the hope of minimising its Gallic character.

2. A new concept of stability

E.C. ZEEMAN

The context is the classification of ordinary differential equations. It is, however, too complicated to try and classify *all* equations, and so we tackle the more modest objective of classifying *generic* equations, for some suitable definite of genericity. It suffices to confine attention to first order differential equations, because any kth order equation, $k > 1$, can be reduced a first order equation on a higher dimensional space by introducing extra variables corresponding to the derivatives of order less than k.

A first order equation on a space X can be written $\dot{x} = v$, where x denotes a point in X, $\dot{x}$ denotes the derivative with respect to time t, and v is a smooth vector field on X. Here X can be a flat manifold such as the real line **R**, or the plane $\mathbf{R}^2$, or n-dimensional euclidean space $\mathbf{R}^n$, or alternatively a curved manifold such as a circle, a sphere, a torus, or any higher dimensional analogue. The easiest space to visualise is the plane $\mathbf{R}^2$ because it can be represented by a piece of paper, and so v can then be represented by drawing little arrows everywhere. The solution of the differential equation is called the *flow*, and this can be represented by drawing the *orbits*, that is the unique smooth curves filling up the plane that are everywhere tangent to v.

The problem of classifying ordinary differential equations is therefore the same as the problem of classifying vector fields on X. So let V denote the space of all smooth vector fields on X. Admittedly V is difficult to visualise because it is infinite dimensional, but nevertheless it is quite easy to handle because it is a vector space, and it has a topology, so we can talk about vector-fields being near one another, and given $v \in V$ we can talk about a neighbourhood of v in V. What is a classification programme? We suggest that it should consist of four steps, as follows.

CLASSIFICATION PROGRAMME

1. Choose an equivalence relation on V, and define a vector field to be *stable* if it has neighbourhood of equivalents in V.

2. Prove that the stables are dense in V.
3. Classify the stable classes.
4. Classify the unstable classes of codimension 1, 2, . . , etc.

Let me explain the programme. If we are going to classify into equivalence classes then obviously we are going to need an equivalence relation: hence Step 1. The choice of equivalence relation will depend upon its usefulness, both its meaningfulness in applications and its fruitfulness in proving theorems; too coarse a relation will lump together things things that 'ought to be distinct', while too fine a relation will separate things that 'ought to be the same'. But the key criterion here will be the notion of stability, not the asymptotic stability of a single orbit but the structural stability of the whole differential equation. Why is stability so important? The answer is that if we are going to use a differential equation to model some application, then the model is bound to be only an approximation, and tomorrow's experiment is bound to be a perturbation of today's, so if our model is going to be of any use in describing the application and predicting tomorrow's experiment then it must be robust under perturbation. In other words it must be stable, so that perturbations of it are equivalent to it; and if it is not stable then we must be able to choose a perturbation that is, which is only possible in general provided the stables are dense in V. Here *dense* means that every open set of V, however small, must contain a stable vector field. Hence Step 2 of our programme. Of course Step 2 is the crunch line behind Step 1: we have to choose an equivalence relation that is sufficiently fine to distinguish things that are qualitatively different, but sufficiently coarse to be able to prove Step 2. Having got over the hurdle of Step 2 we can then tackle Step 3, the business of listing the stable classes.

Finally we come to the business of listing the unstable classes. To visualise what is going on let us pretend for a moment that V is 3-dimensional rather than ∞-dimensional. Let S denote the subset of stables, and T the complementary subset of unstables. Then T will consist of a number of surfaces criss-crossing one other, and dividing up V into a number of regions. The interior of each connected region will be a stable class (or part of a class) and the union of all such interiors will comprise S. Therefore Step 3 means listing the regions. Meanwhile we can divide T into surfaces, edges (where two surfaces cross) and points (where three surfaces cross). Therefore T minus the edges and points will consist of a number of connected pieces of surface, each of which will be an unstable class of codimension 1 (or part of a class). Here *codimension* means the difference between the dimension of V, which we are pretending to be 3, and the dimension of the surface, which we are pretending to be 2. The significance of codimension is as follows. If we are given a single vector field there is no problem: it will be a single point of V and generically it will lie in S; if it happens to lie in T then we can move it into S by an arbitrarily small perturbation. If, however, we are given a 1-parameter family of vector fields then this will be represented by a curve in V, and although we can move such a curve off the edges and points of T it may not be possible to move it off the surfaces of T. Therefore generically it will pierce the surfaces of T, which are the unstable classes of codimension 1. Hence the unstable classes of codimension 1 will classify the generic bifurcations of 1-parameter families of vector fields.

Now look at the edges and points of T. The edges minus the points will consist of a number of connected arcs, each of which will be an unstable class of codimension 2 (or part of a class). It is codimension 2 because we are pretending that V is 3-dimensional and the edge is 1-dimensional. If we are now given a 2-parameter family of vectors fields this will be represented by a surface in V, which we can move off the points of T, but generically will cross the surfaces and edges of T. Therefore the unstable classes of codimension 1 and 2 will classify the generic bifurcations of 2-parameter families. And so on.

In reality V is ∞-dimensional and T will consist of ∞-dimensional submanifolds criss-crossing each other, but nevertheless we can still decompose T into strata of codimension 1, 2, . . . with the property that an r-parameter family will generically only meet those strata of codimension $\leqslant r$. Therefore the unstable classes of codimension $\leqslant r$ will classify the generic bifurcations of r-parameter families.

Summarising: the stable classes classify generic differential equations, and the unstable classes classify their generic bifurcations. Hence the motivation behind the programme.

CLASSIFICATION OF GRADIENT FIELDS

The hope for such a programme was stimulated by the success of Thom's theory of functions and gradient fields. A *gradient field* on X is a vector field of the form $v = -\Delta f$, where $f : X \to R$ is a smooth real function. Let F be the space of all real functions. Choose an equivalence relation on F, as follows: define two functions f, f' to be *equivalent* if they are conjugate, that is to say there exist diffeomorphisms α, β such that the diagram commutes:

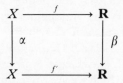

Here a *diffeomorphism* means a one-to-one map such that it and its inverse are differentiable. For simplicity let us assume that the manifold X is compact. Then the programme runs as described above because stable functions are dense in F. A stable function is a Morse function, in other words a function with a finite number of critical points with distinct critical values, each one being a non-degenerate maximum, minimum or saddle. The stable functions are classified by these critical points. The unstable functions are classified by the elementary catastrophes and their Maxwell sets [see 4, 5].

Summarising, the success of the programme for gradient fields lies in describing vector fields by functions, for which there is a complete theory.

STRUCTURAL STABILITY

Thom's success with gradient fields raised hopes for a classification of all vector fields. The favoured equivalence relation was that of topological equivalence, giving rise to the notion of structural stability as follows. Two vector fields v, v' on X are defined to be *topologically equivalent* if there is a homeomorphism of X onto itself throwing v-orbits onto v'-orbits (the orbits of the corresponding differential equations). Here a *homeomorphism* means a one-to-one map such that it and its inverse are continuous. Define a vector field to be *structurally stable* if it has neighbourhood in V of topological equivalents.

There are three criticisms of this definition. Firstly it is inelegant because it lies outside the smooth category, since a homeomorphism need not be differentiable. For example, even if we knew that a given differential equation were topologically equivalent to a standard model, there would not necessarily be any smooth change of coordinates converting the given equation into standard form. Hence topological equivalence is of limited practical use in applications.

Secondly structurally stables are not dense and so there is no guarantee of being able to choose a structurally stable model. Hence structural stability is of limited value in applications.

Thirdly the concept of topological equivalence emphasises the fine structure of orbits, which directs attention away from more important aspects, and which may in fact be irrelevant due to experimental error.

Summarising, the failure of the programme for structural stability lies in describing vector field by orbits, for which there is no complete theory.

STOCHASTIC STABILITY

The new idea that I want to introduce is to describe a vector field by means of a function that measures the asymptotic behaviour of its associated flow. In order to make the function smooth we shall need to introduce a little stochasticity; more precisely we shall choose a small number $\varepsilon > 0$ representing experimental error in the sense that any measurements of order less than ε are meaningless. The intuitive idea of constructing the function is as follows. Given a vector field, solve the associated differential equation, find the attractors of the flow, take the asymptotic measure on those attractors (largest where the flow lingers longest) and then ε-smooth that measure. In practical terms if a computer were asked to draw the flow by starting from a large number of initial conditions and plotting a large number of dots at discrete time intervals, then the function would describe the density of dots, up to order ε.

So much for the intuitive idea: now for the precise definition. Let U denote the space of smooth probability functions on X, that is functions $u:X \to R$ such that $u > 0$ and $\int u = 1$. Given a vector field v the Fokker–Planck equation for v with ε-diffusion is the partial differential equation on X:

$$\frac{\partial u}{\partial t} = \varepsilon \Delta u - \Delta \cdot (uv).$$

Let u denote the steady state given by $\partial u/\partial t = 0$. Then the solution of the Fokker–Planck equation represents a population $u(t)$ being driven along by the vector field v at the same time as being subject to ε-small diffusion, and eventually homing in towards the steady state u, independent of whatever may have been the initial condition. In fact:

Theorem 1. If X is compact then the steady state u exists and is unique, and all solutions of the Fokker–Planck equation tend to u.

We shall give some examples of steady states below, but let us continue with the theory for the moment. We can now use the function u as a tool to study the vector field v. Indeed we have a map $\pi:V \to U$ given by $v \to u$ from the space V of all vector fields to the space U of all probability functions (which is of course dependent upon the parameter ε).

Theorem 2. The map $\pi:V \to U$ is smooth, open and onto.

We can now use π to lift the whole of Thom's classification of functions in U to give a classification of vector fields in V satisfying the four steps of our classification programme. Indeed, it is a strict generalisation of Thom's original classification of gradient fields. In detail, Step 1 of the programme is as follows: define two vector fields v, v' to be *ε-equivalent* if the steady states u, u' are equivalent as functions (as above), and define v to be *ε-stable* if it has a neighbourhood in V of ε-equivalents. It follows from Theorem 2 that a vector field is ε-stable if and only if its steady state is a Morse function. Steps 2, 3 and 4 follow immediately, giving:

Theorem 3. (i) ε-stable vector fields are open dense in V.
(ii) ε-stable classes are classified by Morse functions.
(iii) ε-unstable classes are classified by elementary catastrophes and their Maxwell sets.

I had been trying to develop this idea for several years but had got stuck at establishing the existence of the steady state. What eventually stimulated me into proving it was having to write a paper [6] for the new journal *Nonlinearity*, which I had just persuaded the London Mathematical Society to launch. With the help of the Editor, David Rand, I managed to complete the proof of Theorem 1 just before the Oaxtepec meeting. I was so bubbling over with enthusiasm at the result that the organizers at Oaxtepec kindly allowed me ten minutes to explain it to the conference, even though it did not have any direct applications to biology yet. In those ten minutes I described Theorem 1 and conjectured Theorem 3, and by a happy coincidence Santiago López de Medrano happened to drop in for those ten minutes. By the next day he had sketched a proof of Theorem 2, and so solved my conjectures. The proof of the continuity of π turned out to need a little extra detail, but subsequently he and Marc Chaperon [1] improved this to showing that π was in fact smooth. Together with my student Charlotte Watts we also developed the analogous theory for diffeomorphisms [2, 3].

Now for some examples:

Example 1. Let $X = \mathbf{R}$ and $v = -x$.

The associated differential equation $\dot{x} = -x$ has a point attractor at the origin. It is easy to solve the Fokker–Planck equation in this case, and the steady state turns out to be the normal distribution with mean 0 and variance ε.

Any initial distribution will be driven by the vector field towards the attractor, but will never quite get there because of the diffusion, and so will home in on the normal distribution, which achieves that unique exact balance between attraction and diffusion. Moreover in as much as all differential equations can be thought of as generalisations of this one, the concept of the Fokker–Planck steady state can be thought of as a generalisation of the normal distribution.

Meanwhile this vector field is stable because the steady state is a Morse function. Therefore any sufficiently small perturbation will have an equivalent steady state, again with a single maximum.

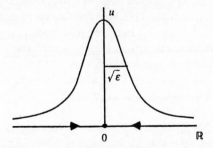

Figure 2.1. The steady state of a point attractor.

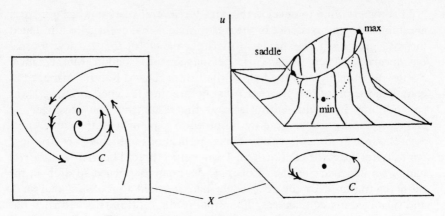

Figure 2.2. The steady state of a cyclic attractor.

Example 2. Let $X = \mathbf{R}^2$, and consider the differential equation

$$\dot{r} = r(1 - r) \qquad \dot{\theta} = 2 - r\cos\theta$$

where (r, θ) are polar coordinates. The resulting flow has a cyclic attractor C on the unit circle and a point repellor at the origin, with all the other orbits flowing towards C. On the attractor itself the flow is slowest where $\theta = 0$ and fastest where $\theta = \pi$. Therefore the steady state u resembles a volcano crater with its rim above the attractor, a maximum above $\theta = 0$, a saddle above $\theta = \pi$, and a minimum above the origin. The vector field is stable because the steady state is Morse.

Example 3. Consider the Van der Pol equation

$$\ddot{y} + K(3y^2 - 1)\dot{y} + y = 0,$$

where K is a large constant. This can be written as a first order equation on the plane:

$$\dot{x} = -y/K \qquad \dot{y} = K(x + y - y^3)$$

where (x, y) are cartesian coordinates. The resulting flow has a cyclic attractor A, a point repellor at the origin, with all the other orbits flowing towards A. On the attractor itself there are two fast vertical legs separated by two slow horizontal legs. Therefore the steady state resembles two parallel mountain ridges with two maxima above the two slow legs, two saddles above two fast legs, a minimum above the origin. The vector field is unstable because by symmetry the two maxima are at the same height, but a small perturbation such as adding a small constant to the equation will stabilise it.

It is interesting that Examples 2 and 3 are topologically equivalent, but not ε-equivalent because the steady states have different numbers of

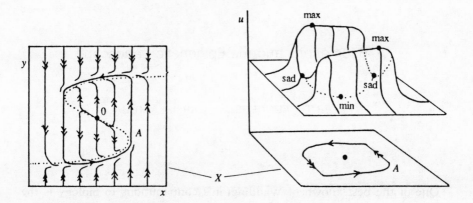

Figure 2.3. The steady state of the Van der Pol equation.

maxima. Indeed in applications the two equations would represent very different phenomena.

Example 4. If X is a compact manifold then most chaotic attractors are ε-stable, because the ε-stables are open dense by Theorem 3. This contrasts strongly with the likelihood that most chaotic attractors are structurally unstable. In effect the ε-smoothing of the steady state hides the fine orbit structure of a chaotic attractor underneath a finite number of peaks [see 7]. Although small perturbations may introduce a pathological orbit structure, such as an infinite number of sinks with tiny basins of attraction, this pathology is irrelevant because it is below experiment error, and indeed will remain hidden beneath the finite number of peaks. Thus the theory may make chaotic behaviour more accessible by bypassing some of the more intractable problems previously associated with chaos.

BIBLIOGRAPHY

1. M. Chaperon and S. López de Medrano, ' "Almost invariant" smooth probability measures for diffeomorphisms and a discrete version of Zeeman's stability theory', Preprint, Ecole polytechnique, 1988.
2. M. Chaperon, S. López de Medrano, C.H. Watts and E.C. Zeeman, 'Almost invariant probability measures for diffeomorphisms and flows on a compact Riemannian manifold, and the associated notion of structural stability', *C.R. Acad. Sci. Paris*, 307 (1988) 95–100.
3. S. López de Medrano, C.H. Watts and E.C. Zeeman, 'Stable diffeomorphisms are dense', Preprint, Warwick University, 1988.
4. R. Thom, *Structural stability and morphogenesis*, Benjamin, New York, 1972.
5. E.C. Zeeman, *Catastrophe theory: selected papers 1972–1977*, Addison-Wesley, New York, 1977.
6. E.C. Zeeman, 'Stability of dynamical systems', *Nonlinearity* 1 (1988) 115–55.
7. E.C. Zeeman, 'On the classification of dynamical systems', *Bull. Lond. Math. Soc.*, 1988 (to appear).

3. Bifurcations and the Epigenetic Landscape

PETER T. SAUNDERS AND CARMELA KUBAL

One of the best known of Waddington's contributions to biology is the 'epigenetic landscape' (Waddington, 1940, see Figure 3.1 below). The developmental system of an organism is portrayed as a mountainous terrain whose shape is determined by guy ropes representing the influence of genes, though, as can be seen from the figure, in a very complicated way. The valleys represent possible pathways along which the development of an organism could in principle take place. A ball rolls down the landscape, and the path it follows indicates the actual developmental process in a particular embryo.

The epigenetic landscape is only a conceptual aid, not a model in the usual sense of the word. It has nevertheless proved very useful, because it illustrates remarkably well many important properties of developing

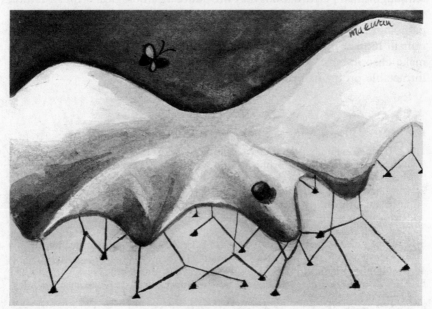

Figure 3.1. The epigenetic landscape. The significance of the butterfly will become clear later.

organisms. For example, if the ball is somehow deflected off its course, then providing it is not pushed over a watershed it will return to the original pathway, though at a point further down the valley rather than to where the disturbance occurred. This corresponds to the property which Waddington called homeorhesis: the ability of the organism to develop normally despite the inevitable perturbations it will experience.

The network of linkages under the surface means that a change in one of the guy ropes will usually have little effect on the shape of the landscape, and what effect it has will depend on its linkages with the other ropes. This illustrates how the effect of any gene is typically dependent on the effects of many others, and that one gene is likely to have an effect on several different properties of the organism. On the other hand, a small change in the landscape can have a major effect on the outcome if it causes the ball to choose a different valley, which is to say that a small genetic change can produce a large phenotypic effect if it causes the organism to follow a different developmental pathway.

Note how this picture of development makes it easy to understand the phenomenon of punctuated equilibria. According to the model, most non-lethal mutations will have little effect, owing to the stability of the pathways. Occasionally, however, a mutation will alter the landscape sufficiently to divert development down a different track, and this will bring about a relatively large change. In general, therefore, large phenotypic changes will not be the mere sum of a sequence of small ones, which is to say that macro- and microevolution are by no means necessarily the same.

As with mutations, most non-lethal perturbations have little influence on the organism on account of the stability of the pathways. However, a sufficiently large perturbation, or one occurring close enough to a fork in the pathway, can divert the ball down a different track. The result will be a change which is not only large, but also very similar to one which is due to a change in the guy ropes. This is the analogue of the phenomenon of phenocopies, in which disturbances during development can bring about the same changes in genetically normal individuals as are observed as the results of mutations.

Waddington was led to the idea of the epigenetic landscape from his knowledge of development. It was a phenomenological model, intended as a conceptual and heuristic aid, and it has been successful in this. As Ho and Saunders (1979) have pointed out, however, we can now see why the epigenetic landscape should have the properties it does. Most models of developmental processes assume that they can be represented by non-linear differential equations. These have a number of interesting and relevant properties which have been known for a long time but which have only recently become familiar to non-specialists. For example, their solutions generally have the same stability property mentioned above, the return to a trajectory rather than to a point. There are frequently multiple solutions and bifurcations. The systems are typically structurally stable, which accounts for the necessary biological property of robustness: most changes in the parameters do not change the qualitative nature of the solutions. Where such changes do occur, their relation to changes in the parameters can be quite complex.

While all the properties of the epigenetic landscape can be understood in terms of those of systems of non-linear differential equations, the converse is not true. It would be astonishing if it were; a pictorial representation, however ingenious, can hardly be expected to capture all the richness of a whole class of complex mathematical structures. The work described in this paper represents the beginning of a project to discover the consequences for development (and hence also for evolution) of the assumption that the processes that are involved can be modelled by non-linear differential equations, making as few hypotheses as possible about the detailed nature of the processes themselves. We are in effect taking a closer look at the topography of the epigenetic landscape. (For more on this use of mathematics in biology, see Saunders, 1989.)

In any general discussion of the properties of non-linear differential equations it is important to bear in mind that most of our intuition about them comes from our experience of systems in the plane, and of course the epigenetic landscape is drawn as a two-dimensional surface. Unfortunately this can be rather misleading, as two dimensions is a rather special case and the behaviour of higher dimensional systems can be quite different. For example, the fact that a closed trajectory divides a two-dimensional space, but not one of higher dimension, into two distinct regions can have significant consequences. Waddington (1957) acknowledged that his model was a low-dimensional representation of a much higher dimensional system, but he did not realize exactly how special a case two dimension is.

Another point is that solutions do not generally bifurcate like a branch railway splitting from a main line. In general, we expect that the original solution will destabilise and that two stable solutions will appear, one on either side (Figure 3.2). If we were to interpret this literally in evolution, it would imply the disappearance of the old species and its replacement by two new ones, but this is too facile. In the first place, we observe such behaviour only if we increase the bifurcation parameter (i.e. the parameter that moves us along the equilibrium trajectory) alone, whereas we shall see that, in general, the situation is seldom so simple. Even if only the bifurcation parameter varies, the phenotype corresponding to one of the new solutions might well be indistinguishable from the old one, or a mutation might alter the parameter in some individuals but not in others. All the same, as we shall also see, studying the details of what happens near bifurcation points can reveal some features which we may expect to observe in evolutionary change.

A valuable tool for studying such problems is catastrophe theory (Thom,

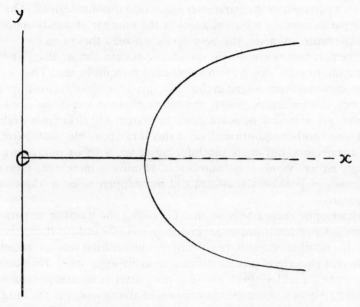

Figure 3.2. A typical bifurcation diagram, with x denoting the bifurcation parameter and y the response. Points on the solid line correspond to stable equilibria, those on the dotted line to unstable equilibria.

1972) because where it is applicable it provides a complete classification of the possibilities. We shall therefore provide a brief explanation of how the theory can be used, and two examples of applications. Primarily, these are to be thought of as illustrations of the method; more detailed analyses of these problems, and of others to which catastrophe theory is not appropriate, are in preparation and will be published elsewhere.

A NOTE ABOUT CATASTROPHE THEORY

Because models of biochemical reaction schemes usually involve some sort of steady-state assumption, catastrophe theory can often be used to predict both some of the phenomena which are likely to be common in development and the way that these are likely to change in evolution.

An introduction to the theory is beyond the scope of this article; the reader who is interested is referred to Saunders (1980) for an elementary account or Poston and Stewart (1977) for a more sophisticated version. The following brief description is included, however, in the hope of making the next two sections comprehensible to those not familiar with catastrophe theory.

Non-linear differential equations, and a fortiori systems of such equations, typically have multiple steady states. In general, the number of steady states will not be the same for all values of the parameters, and this implies a partition of the parameter space into disjoint regions. The values of the parameters for which changes in the number of steady states occur form the bifurcation set, the boundaries between these regions.

If the parameters are slowly varied, the system almost always responds by altering its state slowly, remaining close to equilibrium. This is exactly the process that is envisaged in the usual quasi-steady-state assumption. If, however, the trajectory crosses the bifurcation set and if the equilibrium that the system is in disappears, then the system will either move relatively quickly to another equilibrium, or, if this is not possible, destabilize. Note that a trajectory can cross the bifurcation set without inducing a rapid change, either because the number of equilibria increased, rather than decreased, or because the system did not happen to be in the state that disappeared.

Castastrophe theory tells us that providing the number of parameters that are independently important in determining the sudden change is not too great, the number of qualitatively different bifurcation sets (i.e. the number of different patterns of discontinuities) is surprisingly small. No matter how many state variables there are (i.e. no matter how many variables are required to give a complete description of the system), if the number of relevant parameters is no greater than five, the number of state variables which have to be taken into account is at most two, and the number of distinct 'elementary catastrophes' is eleven. Each of these has the same bifurcation behaviour as a gradient system with a canonical form which is relatively straightforward to derive and analyse (for details, see either Saunders, 1980, or Poston and Stewart, 1977).

The restriction to five 'control variables' is less restrictive than it seems, because if a system is exhibiting abrupt changes, and if more than five parameters are crucially involved, it is going to be very difficult to make sense of the system by any means. Moreover, while the higher-order catastrophes are more difficult to analyse, they are by no means intractable.

It should be noted that neither the five control variables nor the one or two essential state variables are necessarily quantities that are directly measurable. In general, they are functions of observable control and state variables, respectively; we may think of them rather like eigenvariables. As we shall see, this makes the task of applying the theory more complicated than it would otherwise be, but not impossibly so.

THE CUSP CATASTROPHE

A number of authors have studied chemical and biochemical reactions which exhibit multiple steady states and sharp transitions between them. The systems they describe tend to have much in common, as indeed catastrophe theory suggests they must; as an illustration, therefore, we outline only one, that of Lewis, Slack and Wolpert (1977), which was proposed in connection with developmental biology and is therefore directly relevant to the present work.

The aim of the model was to show how discrete states could arise out of continuous gradients. The states were supposed to be specified by the concentration of a gene product denoted by g. The gene was activated by a 'signal substance' S, and the rate of change was given by the equation:

$$\frac{dg}{dt} = K_1 S + \frac{K_2 g^2}{K_3 + g^2} - K_4 g \tag{3.1}$$

where the K_i are all constants.

Lewis et al. set K_2 and K_3 equal to unity and K_4 equal to 0.4. Without loss of generality we can choose the units of S such that K_1 is also equal to unity, and it is then comparatively straightforward to see how boundaries can arise.

Suppose that both g and S are initially zero and that S is then slowly increased. Then dg/dt will be positive, so g will increase as well. Because dg/dt will be small, g will always be close to the equilibrium value (for the current value of S) which can be found by solving the equation $dg/dt = 0$ for g.

With the given values for the constants, this equation is:

$$2g^3 - 5g^2(1 + S) + 2g - 5S = 0 \tag{3.2}$$

and for $S = 0$ there are three real roots: 0, 0.5 and 2. There are thus three possible steady states, and it is not hard to show that those at $g = 0$ and $g = 2$ are stable and that at $g = 0.5$ is unstable. Consequently, if g is initially zero it will remain so until S begins to increase. Even then, the situation will be much the same for a while in that there will still be two stable steady states with an unstable one in between. The smallest equilibrium value of g will be positive, so as S increases slowly, so will g, remaining at all times close to equilibrium.

If, however, S is increased above a critical value S_c, which is approximately 0.041, a significant change occurs. Equation (3.2) now has only one real root, together with a complex conjugate pair. The two steady states (one stable, one unstable) corresponding to smaller values of g have coalesced and disappeared. The system will therefore move rapidly to the higher equilibrium, i.e. there will be a rapid increase in g. Thus, at this point, a very small change in S, from just below S_c to just above it, will cause a large change in g. Consequently, if there is a smooth gradient in S throughout a region, at the position where S takes on the value S_c, there will be a discontinuity in g, i.e. a sharp frontier between two sub-regions.

Saunders and Ho (1985) used catastrophe theory to extend this result in two important ways. First, while the fixing of the constants K_i still permitted Lewis et al. to demonstrate the formation of boundaries, which was

what they had set out to do, it concealed other important properties of
Equation (3.1). What is more, the theory tells us that these properties are
not peculiar to this equation. They are likely to be found in almost any
system which is capable of producing a transition from one steady state to
another, and hence of producing a boundary in the way that this model
does. Consequently, once we observe a transition from one steady state to
another, we ought to look for the other phenomena too; they are not
artefacts of a particular mathematical model.

The crucial point is that the condition for equilibrium, Equation (3.2),
is a cubic, as it is for cusp catastrophe. The canonical form of the cusp (i.e.
the potential of the canonical gradient system with the same bifurcation
behaviour) is, however:

$$V(x) = x^4 + ux^2 + vx \qquad (3.3)$$

The equilibrium condition is, accordingly:

$$0 = dV/dx = 4x^3 + 2ux + v \qquad (3.4)$$

and both these have two parameters. This tells us that there is more to the
model than Lewis *et al.* found, that a second parameter must be allowed to
vary. It also tells us that varying one more parameter will be sufficient, as
the model will then be structurally stable. Saunders and Ho chose K_3
(which they wrote simply as K, because the other K_i were set equal to unity)
because varying a saturation constant seems physically plausible, but
allowing the other parameters to vary as well would not have changed the
general pattern of the behaviour, only the details.

The equilibrium condition is now:

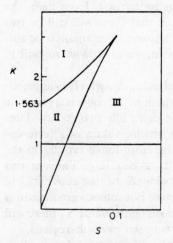

Figure 3.3. The control space for equation (5), showing the bifurcation set. Equilibrium is
possible only at a low value of g in region I and at a high value in region III; in region II there
are two stable equilibria. The line $K = 1$ corresponds to the model of Lewis *et al.* (1977) with
an abrupt change occurring at $S = 0.041$. (From Saunders and Ho, 1985.)

$$2g^3 - 5g^2(1 + S) + 2Kg - 5KS = 0 \qquad (3.5)$$

For our purposes this is equivalent to Equation (3.4), even though g, S, K are not the canonical variables x, u, v but functions of them.

A full account of the analysis of Equation (3.5) is given in Saunders and Ho (1985) but the results can be summarized as follows. Figure 3.3 shows the control space for the system, with the bifurcation set indicated. If we assume, as before, that initially both g and S are zero and that S is then slowly increased, then as the trajectory crosses the right-hand branch of the cusp the value of g will move rapidly to a higher value. If S is then decreased, the change back to a low value will occur as the trajectory crosses the left branch of the cusp, which for $K < 1.563$ would require S to be negative. Hence, for all values of K below 1.563, even if S returns to zero, the concentration of the gene product g will remain at a significantly non-zero value.

To see how a system of this kind might evolve, consider the effect of a change in the saturation constant K. Suppose that, initially, K is small. Then, if S increases to about 0.03, say, g will reach a high equilibrium value. Now suppose that, in evolution, K increases to about 1.4, say. Then, for the same S, g will stabilise at a low value. But if S is perturbed to a higher value, g will move to a high value which it will remain at, even if S subsequently falls back to 0.03. Finally, if K increases even further, to much over 2 say, then the system is insensitive to a perturbation in S and will return to the low equilibrium value of g.

What happens in the other direction? Suppose the original form is $K > 2$, $S = 0.03$. Then if we reduce K to 1.4 there is no change; we still get the low value. It can be perturbed to the high value but that is as before; we can still only phenocopy in one direction. So a prediction of this model is that if it is possible to produce a phenocopy of a mutant organism from an individual which is genetically normal, it will not be possible to produce a phenocopy of a normal organism from a mutant. In terms of the epigenetic landscape, it is likely that there are watersheds which can be crossed in one direction but not in the other, which is difficult to represent on a picture. (Lewis *et al.*, even though they allowed only S to vary, also concluded that change would occur in one direction only.)

If K changes gradually in evolution, so that it passes through a number of intermediate values, then there will be a period during which it is possible to phenocopy the old form (or the one that is about to appear). After a while this ability will disappear or (in the latter case) the new form will appear as a mutant and it will not be possible to phenocopy the old one.

It is, of course, only because we have a particular model in mind that we are able to make definite statements about under what conditions certain things will happen. But our results also tell us what we may expect from situations in which we do not know the mechanism and cannot therefore write down the equations. If there is a transition from one state to another, then there is likely to be a range of control variables for which both equilibria are possible. We would therefore expect hysteresis, and while the effect might not be so striking as above (though it could be) it would still imply a greater stability of the transformed state than would otherwise be expected. We would also expect that phenocopying would be possible for some values of the control parameters but not others, and that it would typically be possible in one direction only.

THE BUTTERFLY

Applications of catastrophe theory to the study of complex systems usually involve the assumption that the mechanism is the simplest that is consistent with the observed behaviour. An important advantage of the theory is that it allows us to say in a straightforward way what we mean by 'simplest'; when we are dealing with a transition from one state to another it means one that leads to the cusp catastrophe, which has codimension 2 (i.e. two control variables). There are three catastrophes of codimension 3 (the swallowtail and the elliptic and hyperbolic umbilics), but the next one of interest here is the butterfly, which is of codimension 4 and has the canonical form:

$$V(x) = x^6 + tx^4 + ux^3 + vx^2 + wx \qquad (3.6)$$

Because the parameter space is four-dimensional, it is rather harder than before to see what is going on. The best way is to consider what happens in the (v, w) plane for various fixed values of t and u. It is relatively easy to show (see, e.g., Saunders, 1980, pp. 52ff) that there are two distinct possibilities. If

$$5u^2 + (4t/3)^3 > 0 \qquad (3.7)$$

then the projection of the bifurcation set into the (v, w) plane is a cusp, though, in general, it is distorted from the standard form. If, however, $5u^2 + (4t/3)^3$ is negative, then a new feature appears, a pocket within which there are five equilibria, three stable and two unstable (Figure 3.4). To assist in predicting when rapid changes will occur, we have marked on the diagram which stable equilibria are available in each of the regions into which the (v, w) plane is divided.

We have already seen that the canonical control variables of a catastrophe are not, in general, the variables that we observe directly. This makes it difficult to predict what will happen in general terms, partly because we do not know which trajectories are physically reasonable (and in any case, the trajectories are now in four dimensions, not two) and partly because of the importance of constraints on the observable parameters (e.g. that they all be positive) which transform into less obvious conditions on the ranges of the canonical variables.

To see the sort of thing that can happen, we consider a system like that

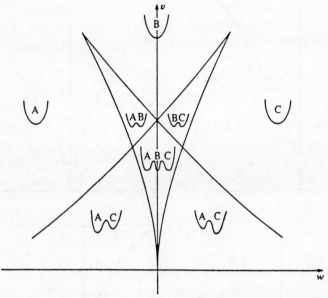

Figure 3.4. The projection into the (v, w) plane of the bifurcation set of the canonical butterfly catastrophe with u = 0 and t < 0. The form of the potential V(x) [equation (6)] in each of the regions is shown, and the minima have been labelled to assist in predicting what will happen as the control trajectory crosses different parts of the bifurcation set. (After Saunders, 1980.)

analysed in the preceding section but this time with two separate quadratic saturation terms:

$$\frac{dg}{dt} = S + \frac{ag^2}{K + g^2} + \frac{bg^2}{L + g^2} - Dg \qquad (3.8)$$

We have chosen this model because of its similarity to the previous one rather than as a description of any particular process in development. We note, however, that equations with two Michaelis-Menten terms are not uncommon in the literature, so Equation (3.8) is at least a plausible example.

The condition for equilibrium, the analogue of Equation (3.5), is:

$$Dg^5 - (a + b + S)g^4 + D(K + L)g^3 - (S(K + L) + bK + aL)g^2$$
$$+ DKLg - SKL = 0 \qquad (3.9)$$

The easiest way to analyse the bifurcation behaviour of the system is to put this equation into the standard form by the co-ordinate transformation $x = g - (a + b + S)/5D$. The canonical variables t, u, v, w will of course be rather complicated functions of the observable variables a, b, K, L, D, S. It is therefore not possible to give a simple account of what happens for all possible control trajectories, but we summarise here what appear to us to be the most significant features.

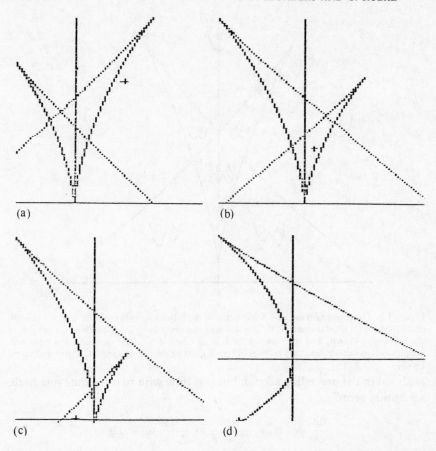

Figure 3.5. The projection into the (v, w) plane of the bifurcation set of the canonical butterfly catastrophe for values of u and t corresponding to K = 36.6, L = 0.75, A = 1, B = 0.14, D = 0.11, and (a) S = 0, (b) S = 0.003, (c) S = 0.005, (d) S = 0.01, (e) S = 0.03. The position of the control point is also shown. The scale is not the same for each figure.

For most combinations of parameter values, the projection of the bifurcation set into the (v, w) plane is a simple cusp and the model is very much like the previous one. For others, however, it is more complicated. If, for example, we take K = 36.6, A = 1, B = 0.14, D = 0.11 and L greater than zero but less than about 2.5, then the pocket is present for small values of S and the behaviour can therefore be quite different.

To follow the trajectory of the system as S is varied is more difficult than before, because changing S changes all the canonical variables, not just v and w. Consequently, the projection of the bifurcation set also changes, i.e. as the point representing the values of the control variables v and w moves, the figure in the plane that marks the location of the possible jumps changes as well. Fortunately, this does not affect the analysis for the parameter values we considered because the qualitative changes in the

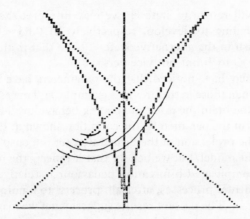

Figure 3.6. Control trajectories in the (v, w) plane for (top to bottom) L = 0.55, 0.6, 0.65, 0.75. The diagram is to be interpreted as described in the text.

bifurcation set do not affect the region in which the control point is (see Figure 3.5.) We can then describe what happens as though Figure 3.4 remained unaltered throughout. We may think of this as a projection of the trajectory on to the (v, w) plane in which the line joining a point to its projection need not be orthogonal to the plane nor even a straight line, but must not cross the bifurcation set.

Figure 3.6 shows the projections of some trajectories as S increases from zero to about 0.5 for different values of L. In all cases, for S sufficiently large, the trajectory passes into the region in which the only steady state is C, which corresponds to large values of g. On the other hand, the point $S = 0$ is in different regions for different values of L.

To see how this model can differ from the previous one, suppose that S and g are both initially zero and that S is then increased to 0.5 and allowed to fall to zero again. In all cases shown, g will move from the lowest equilibrium A to the highest one C, but the final state will depend on the value of L. If $L = 0.75$ it will return to A, if $L = 0.65$ it will finish in state B, whereas if $L = 0.6$ or 0.55 it will remain in C. Where $L = 0.6$, if the system is perturbed to state B it will remain there, whereas if $L = 0.55$ it will return to C. If the system would normally finish in state B ($L = 0.65$) it cannot be perturbed into state C. If we interpret state A as undetermined and B and C as two alternative determined states, then once again we find that phenocopying is possible in one direction only.

Another possibility is that S is increased to 0.5 and maintained at that value, i.e. the signal substance builds up to a significant final value rather than appearing as a transient pulse. In that case, the transition will be to C for all positive values of L less than about 2.5. Now suppose that a disturbance destroys S, so that its concentration falls to zero. Then for L less than about 0.63 the system will remain in state C, whereas for L greater than about 0.72 it will return to state A; we may interpret these as surviving unaffected or failing to develop, respectively. If $0.63 < L < 0.72$, the system will finish in the alternative state b. Note that in this scenario too, we can perturb C to B but not vice versa.

The reader may have noticed that the parameters have been given less natural values than those in the previous example, and may suspect that this was necessary to obtain the more interesting behaviour. This is indeed so: for most values of the parameters, the condition shown in Equation (3.7) is satisfied, and the projection of the bifurcation set is a cusp. This might be an artefact of the model, but we believe that it reflects the real situation; in other words, complicated bifurcation behaviour is relatively uncommon. Most developmental processes, after all, proceed to a unique end-point if they proceed at all. The present results suggest what we may observe in the relatively few situations in which there are more possibilities. Of course such situations, though few in number, are likely to be very significant when they do occur.

In their studies of elastic stability, Thompson and Hunt (1973) demonstrated that when two failure modes coincide, the result is more complicated (and, in the case of real structures, more dangerous) than might have been anticipated. We expect that in much the same way, complex bifurcation behaviour will be observed when the bifurcations associated with two separate reactions occur more or less together. For example, we would expect that, in a sequence of reactions, the most interesting behaviour will occur when no single step can be identified as the sole limiting reaction. One could imagine that if a parameter is changing gradually, either in development or in evolution, this might be true only for a relatively short period.

CONCLUSIONS

The results presented here are preliminary, intended chiefly to serve as illustrations and to demonstrate that the programme we are proposing can actually be put into practice. We have nevertheless found one feature of the systems we have studied that is not captured by the epigenetic landscape – namely, that diversion to an alternative steady state is typically possible in one direction only. On the basis of what we have discovered so far, we would not expect that one could produce a phenocopied normal *Drosophila* from an individual that was genetically *bithorax*. On an evolutionary time-scale, the result suggests a source of irreversibility, and consequently of the tendency towards increase in complexity (cf. Saunders and Ho, 1986).

Our use of words such as expect and suggest, rather than predict, is deliberate. For even if we are able to strengthen our result mathematically, this will not establish it as a firm law of development. The great complexity of the developmental system implies that there can be situations to which general results of this kind do not apply. However large the class of mechanisms covered by the theorems, there are bound to be some that are not included. In particular, while the assumption that development proceeds as simply as possible is a good working hypothesis, it is not always true, and even where it is we may interpret it incorrectly if we are working at the wrong level. The process we are investigating may be more appropriately studied as part of a larger one, and what is simpler globally may well appear more complicated locally. For example, a system with two separate bifurcations could exhibit phenocopying in both directions even if our result held for each bifurcation separately.

What we are trying to do is to discover certain features of organisms and their evolution which we expect to occur frequently, to be typical. There will, however, be exceptions. Indeed, the exceptions will be of special interest because they will highlight those aspects of development or evolution which do not appear to arise naturally out of simple processes and which therefore require special explanation. Moreover, while results that admit exceptions may appear unsatisfactory to those accustomed to the rigorous laws of physics, they are appropriate to biology. Consider, for example, the so-called laws of evolution, such as Cope's Rule (that there is a trend towards increase in body size), Williston's Law (that there is a reduction in number of identical parts and a specialization of those that remain), Dollo's Law (that a feature, once lost, is unlikely to reappear), etc. There are exceptions to all of these, yet the regularities they describe are considered by palaeontologists to be common enough and significant enough that it is useful to identify them and give them names. If this is true of the observations, then we may expect it to be true of our theoretical work as well (see Saunders and Ho, 1986; also Rescher, 1970, Ch. X).

Waddington designed the epigenetic landscape to illustrate and highlight certain properties of development which he considered important. Our aim is to extend his idea and so convert it into a theoretical tool for the study of development and evolution. In doing this, we have to be mindful of the sort of role which theory can play in these subjects and in biology in general. If we want to derive theoretical results in a science which is less exact than physics we have to use methods that recognise the difference. It may be appropriate to clamp a stone in a vice in order to study it, but if we want to learn about butterflies we have to hold them gently.

ACKNOWLEDGEMENT

We are grateful to Mae-Wan Ho for helpful comments and for drawing Figure 3.1.

REFERENCES

Ho, M.W. and Saunders, P.T. (1979) Beyond neo-Darwinism: an epigenetic approach to evolution. *J. Theor. Biol. 78*, 573–91.

Lewis, J., Slack, J.M. and Wolpert, L. (1977) Thresholds in development. *J. Theor. Biol. 65*, 579–90.

Poston, T. and Stewart, I.N. (1977) *Catastrophe Theory and its Applications*. London: Pitman.

Rescher, N. (1970) *Scientific Explanation*. New York: The Free Press.

Saunders, P.T. (1980) *An Introduction to Catastrophe Theory*. Cambridge: Cambridge University Press.

—— (1989) Mathematics, structuralism and the formal cause in biology. In B.C. Goodwin, A. Sibatani and G. Webster (eds.) *Dynamic Structures in Biology*. Edinburgh: Edinburgh University Press (in press).

Saunders, P.T. and Ho, M.W. (1985) Primary and secondary waves in prepattern formation. *J. Theor. Biol. 114*, 491–504.

—— (1986) Thermodynamics and complex systems. In C.W. Kilmister (ed.) *Disequilibrium and Self-Organisation*. Dordrecht: Reidel.

Thom, R. (1972) *Stabilité Structurelle et Morphogénèse*. Reading: Benjamin.

Thompson, J.M.T. and Hunt, G.W. (1973) *A General Theory of Elastic Stability*. London: Wiley.

Waddington, C.H. (1940) *Organisers and Genes*. Cambridge: Cambridge University Press.

—— (1957) *The Strategy of the Genes*. London: George Allen & Unwin.

4. Pattern formation and pattern selection in reaction–diffusion systems

Mathematical biologists have always been fascinated by the problem of pattern formation and pattern selection. In 1952, Turing suggested that reaction–diffusion equations may be a suitable mechanism to explain some of the patterns that arise in biology, particularly in morphogenesis. Since then, this idea has been studied extensively. The book by Meinhardt (1982) is a valuable reference: more recent work includes work by Murray and co-workers (Arcuri and Murray, 1986), De Dier and others (De Dier *et al.*, 1987a,b,c), and the present author (Eilbeck, 1983, 1987, 1988; Duncan and Eilbeck 1988). A number of papers in this area will be found in the proceedings of the 1987 Capri conference, edited by Ricciardi (1988).

Recently, Murray and others have extended the reaction–diffusion model to take into account mechanical forces in membranes or between cells (Lane *et al.*, 1987). We shall not consider this extension here.

One aim of this paper is to survey what can be done with modern numerical methods in studying systems of reaction–diffusion equations. We shall not discuss the details of these methods here, since they are described in some of the above references, but we shall attempt a deliberately non-technical survey of the results of such investigations. This paper makes particular use of the studies of Duncan and Eilbeck (1988) and Eilbeck (1988).

Suppose that you wish to explain the generation of a pattern or sequence of patterns in a model biological or ecological system, and you hope to model this by one or more coupled reaction–diffusion equations. To be specific, we consider two coupled reaction–diffusion equations of the form

$$u_t(x,t) = d_1 \Delta u(x,t) + F(u,v) \tag{4.1}$$

$$v_t(x,t) = d_2 \Delta v(x,t) + G(u,v)$$

Here $x \in \Omega \subset \mathbb{R}^n$, $n = 1$, 2 or 3 (i.e. we are solving the problem on some bounded region in one, two or three space dimensions), the time variable is $t \geqq 0$, and d_1 and d_2 are real diffusion coefficients. The functions F and G are non-linear functions of u and v and may also depend on various parameters. The dependent variables u and v satisfy some sort of boundary conditions on $\partial\Omega$, the boundary of Ω. Again, to be specific we shall assume in this paper that these conditions are Neumann boundary conditions:

$$\frac{\partial u}{\partial n} = \frac{\partial v}{\partial n} = 0 \qquad (4.2)$$

where $\partial/\partial n$ is the normal derivative of $\partial\Omega$. This choice makes some of the analysis simpler, but the general techniques we describe carry over to other choices of boundary conditions.

The patterns generated by solutions of such equations may evolve as a time-dependent problem starting from some initial distribution of morphogens or other chemicals, populations, etc., and ending up as some other steady-state or time-dependent pattern. In this case we would consider the parameters of the model as fixed – the final configuration depends only on the initial configuration and the chosen values of the parameters. In other cases we may imagine the parameters of the model as changing in time, and the observed patterns will depend on the initial conditions and on the particular choice of how the parameters vary. We may get different results if the parameters vary quickly with time instead of slowly.

One common model which involves parameters varying in time is to imagine that the system we are modeling is growing. The size of the system will hence be a time-dependent parameter, and it is well known that different sizes of systems will support different patterns. Alternatively, we can consider systems in which the diffusion rates vary in time, due for example to the growth of intra-cellular membranes. By a simple rescaling argument, it is easy to show that this is equivalent in some cases to varying the size of the system. In this paper we consider such a simple growth model. We also assume that the growth in time is much slower than the diffusion process. This means that, if for fixed size the system evolves to a steady-state solution, for varying times the system will follow a path of stable steady-state solutions in the parameter space. If at any value of the parameter the current pattern becomes unstable, it evolves quickly to another stable pattern. This assumption simplifies the calculations, but can be dispensed with if a more precise model is appropriate.

Consider the one-dimensional case for simplicity. If the length of the system is $2L$, i.e. $x \in [-L, L]$, we can scale this on to the unit interval $[0,1]$. If we consider steady-state solutions ($u_t = v_t = 0$), we can simply incorporate this scaling into (4.1) to give the equations:

$$d_1 \Delta u(x,t) + \alpha F(u,v) = 0 \qquad (4.3)$$

$$d_2 \Delta v(x,t) + \alpha G(u,v) = 0$$

where $\alpha = 4L^2$. We could if we wish carry out a further scaling to make one of the diffusion coefficients equal to unity: we shall not do this, but it should be borne in mind that only the value of the ratio d_1/d_2 is significant.

We will consider only one specific model in this paper for F and G, the so-called Sel'kov model (Sel'kov, 1968):

$$F(u,v) = 1 - uv^p, \; G(u,v) = c(uv^p - v) \qquad (4.4)$$

with $p = 3$ and $c = 0.25$. However, experience with other reaction schemes suggests the qualitative results are generic for other inhibition–activation models.

In the next section of this paper we present some results on one-dimensional regions, with some brief comments on the possibility of varying phase model in these cases. In the last section we briefly review some results in two space dimensions.

ONE-DIMENSIONAL RESULTS

Although the one-dimensional case is somewhat unphysical, understanding the results here can greatly simplify the understanding of more complicated higher-dimensional cases.

As a first stage in the analysis, we will need a 'map' of all the steady-state solutions available to us in the parameter region of interest, or in mathematical language, a bifurcation diagram. Such diagrams can now be routinely calculated (cf. the references by the present author and by De Dier *et al.*). In such a diagram, we plot in some schematic way the number of solutions as a function of a chosen parameter called the primary bifurcation parameter. All other (secondary) parameters are held fixed. For the model considered here, our primary parameter is α, related ($= 4L^2$) to the size of the system. The ratio d_1/d_2 is regarded as fixed: different values for d_1/d_2 will give different bifurcation diagrams and a different sequence of patterns as the system grows.

In particular, if d_1/d_2 is small enough, it is straightforward to show that the homogeneous solution (in this case $u = v = 1.0$) is stable for all α (Brown and Eilbeck, 1982). If we increase the size of d_1/d_2 we will eventually reach a value for which the homogeneous solution becomes unstable for some values of α. Figure 4.1 shows such a bifurcation diagram for $d_1/d_2 = 0.3/0.0384$. In this figure we have plotted the value of one component, u, at the left-hand edge of the region, as function of $\ell n\, \alpha$, rather than α. The choice of a log plot for the α axis leads to some simplification, as discussed below. We see that for some values of α, only the constant solution $u = 1$ is a possible solution. However, for other α values there are three or more solutions. One of these is the constant solution, whereas the other solutions are inhomogeneous in space. Remember that each *point* on the bifurcation diagram corresponds to a single *function* $u(x)$ (look ahead to Figure 4.2 for a range of function plots for different L values). In Figure 4.1, solutions on the bifurcation plot which are stable to (small) perturbations are drawn as solid lines, whereas solutions which are unstable are drawn as dashed lines.

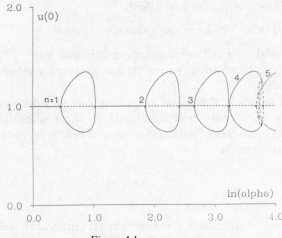

Figure 4.1.

Points at which the number of solutions change are bifurcation points. Each curve of inhomogeneous solutions is labelled by an integer n: this integer describes the shape of the solution near the constant solution, since in this region $u(x) - 1 \sim \pm \cos n\pi x$. However, this labelling convenience should not detract from the fact that, away from the constant solution, many Fourier modes may be present in the solution. However there is often some symmetry which is maintained: for example the curve of solutions described by $n = 2$ gives solutions which are always symmetric about the midpoint of the region (i.e contains only even harmonics).

Each curve of non-homogeneous solutions is double valued: the reasons for this are straightforward. A solution which contains some odd harmonics can be used to obtain a new solution which is just its mirror image: $u(x) \to u(1 - x)$. This new solution will have a different value of $u(0)$ and will appear to be a different solution in the bifurcation diagram. Solutions which contain only even harmonics can be used to generate other solutions in a rather different way, as described in Duncan and Eilbeck (1988) and Eilbeck (1988).

The advantages of plotting the diagram on a log scale in α are clear from inspection: each curve $n = 2, 3,...$ appears on the diagram as a simple copy of the $n = 1$ curve, merely translated to the right. The reason for this depends on simple symmetry arguments and the fact that we have imposed no-flux boundary conditions (4.2); for details, see Duncan and Eilbeck (1988) and Eilbeck (1988).

Now consider the range of patterns that the system can take as L increases from a small value. One possible sequence is shown in Figure 4.2, which shows a plot of $v(x)$ against x for various discrete values of L. Note the sequence is not unique: at the first bifurcation point at $\alpha = 1.5995$, the constant solution becomes unstable, and the system will adopt a pattern whose deviation from the constant solution is proportional to $\pm \cos \pi x$. The

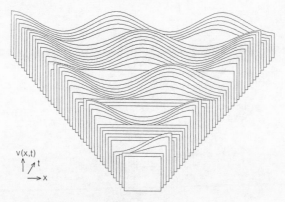

Figure 4.2.

choice of sign is arbitrary and different small perturbations, of the sort arising in any physical situation, will nudge the solution into one or other of these states with equal probability. To insist on one particular polarity for the pattern, we must impose a specific asymmetrical disturbance which will break the $\pm$ symmetry. One possibility occurring in embryology is the act of fertilisation, a large asymmetric perturbation which presumably fixes the orientation of the first pattern. However, this event happens only once, and in order to follow the sequence shown in Figure 4.2 we must make further symmetry breaking choices at the points where the $n = 2,3$ and four curves branch from the constant solution. One possible explanation is that the first pattern sets up a secondary permanent pattern which in turn reacts with the system considered here in a complicated way to fix the sequence of higher patterns.

Note the large regions of α (and L) above the first $n = 1$ branch for which no pattern exists. However these regions become smaller and smaller as L

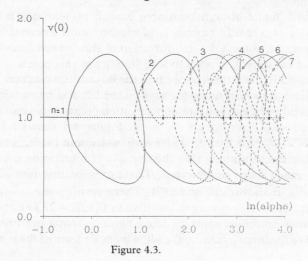

Figure 4.3.

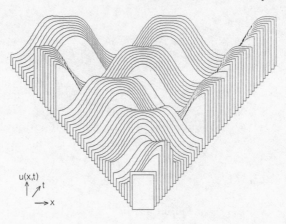

u(x,t)

Figure 4.4.

increases, almost vanishing between the $n = 3$ and $n = 4$ branches, and vanishing completely above $n = 4$. In fact the $n = 4$ branch becomes unstable before it returns to the constant solution. To follow the solution from this point requires a numerical integration of the full time-dependent equations (4.1). These numerical studies show that the solution evolves to the stable $n = 5$ branch, although which half of the branch is chosen depends again on the nature of the small perturbation required to shift the solution from its unstable equilibrium point. The $n = 4$ solution contains harmonics $n = 8,12,...$ whereas the $n = 5$ solution contains harmonics having $n = 10,15,...$. The amplitude of these harmonics decrease rapidly with increasing n, and the common harmonic at $n = 20$ is not strong enough to influence the choice of switch given reasonably sized perturbation. Once again, an external space-dependent field is required to make this choice unique.

The small 'figure-of-eight' unstable branch of solutions between the $n = 4$ and $n = 5$ main branches is of mathematical interest, but also of physical interest, since it is the bifurcation of this branch from the $n = 4$ branch that causes that branch to lose stability at this point.

As we increase the ratio d_1/d_2 we find the bifurcation diagram Figure 4.1 changes. Each n branch tends to move to the left and eventually overlaps with its neighbours. As each pair of bifurcation points cross, more and more complicated subsidiary curves are born. Figure 4.3 shows a bifurcation diagram for $d_1/d_2 = 0.3/0.033$ (redrawn from Eilbeck 1988). Now there are no stable constant solutions after the first $n = 1$ bifurcation point, and the picture is much more complicated. The corresponding plot of $u(x)$ as a function of L is shown in Figure 4.4. There is still some arbitrariness in calculating Figure 4.4, but much less than in Figure 4.2. The choice of the $n = 1$ branch to follow must be fixed by an external perturbation, but either half of the branch leads to the same biopolar form of the $n = 2$ curve.

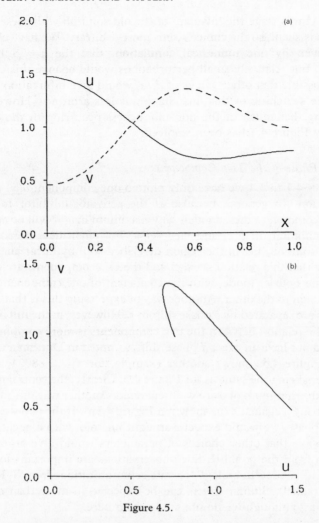

Figure 4.5.

Once this branch loses stability at $\alpha \sim 11.9$, it switches to a specific $n = 3$ branch. The reasons that the choice appears to be unique in this case is that both branches share a common harmonic, at $n = 6$, which has a reasonably large amplitude. This is sufficient to nudge the unstable $n = 2$ solution in a specific direction, independent of any small external perturbations. In the language of quantum mechanics, we would say that the two solutions have a large overlap integral. Note that the overlap would be zero if we were making a linear approximation, and no higher harmonics were present. When this branch loses stability at $\alpha \sim 21.1$, the solution follows for a short while an asymmetric secondary branch, then switches to a specific $n = 4$ branch for the same reason as above. Once the $n = 4$ branch becomes unstable, the solution can now switch to either an $n = 5$ or an $n = 6$

branch. At this stage the 'overlap' of the old solution with these two new solutions is small, so the choice seems more arbitrary. We have chosen the result given by one numerical simulation, that the $n = 5$ branch is favoured, but relatively small perturbations would change this choice.

It is possible that other values of d_1/d_2 would give bifurcation diagrams where the switching of this final stage was less arbitrary. However, it is interesting that much of the non-uniqueness occurring in the case described by Figure 4.1 has been resolved.

Relative Phase of the Two Components

In Figures 4.1 to 4.4 we have only plotted one component $u(x)$ or $v(x)$ of the solution. In general, because of the activator-inhibitor mechanism which is common to these models, any maximum in $u(x)$ will be approximately a minimum in $v(x)$ and vice versa. For small deviations from the constant solution (i.e. in the region described well by linear analysis), we can show that this relation is exact and the two components are described by a single Fourier mode, with the coefficient of one component being of different sign to the other. Another way of expressing this is that the linear solutions are separated by a phase shift of π. However, in the full non-linear region, the relation between the two components is more complicated and they need not have an exact π phase shift, as noted in Duncan and Eilbeck (1988). Figure 4.5a shows another example, for $\alpha = 2.88024$, the other parameters being the same as for Figure 4.3. Clearly, the central maximum of u and the minimum of v are well separated. Another way of showing this is by plotting u against v (as shown in Figure 4.5b). If the two were exactly π out of phase, we would expect a straight line plot with a negative slope. It is possible that other choices of parameters would give an even large deviation from the 'π shift'. Such observations are important for various 'phase field' models in biology (private communications from S. Kauffman and J. Frankel). Similar effects can be expected in more than one space dimensions, although we do not study them here.

TWO SPACE DIMENSIONS

In more than one space dimension the patterns formed by reaction–diffusion equations can obviously be more diverse, and not surprisingly the numerical calculations required to analyse them require more thought and a bigger computing budget. The pseudo-spectral method gives an efficient technique for tackling these problems, but there are still some interesting unsolved problems concerning the optimum choice of collocation points. To date, very little has been done on fully three-dimensional problems, despite the fact that these are of most obvious biological interest. In the next few years such calculations will provide an interesting area of study.

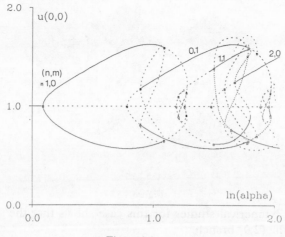

Figure 4.6.

In this paper we consider only two-dimensional studies, in particular the study of solutions in a rectangular region carried in our earlier studies (Eilbeck, 1988). See also the introductory investigation into axially symmetric solutions inside a sphere described in Duncan and Eilbeck (1988).

The bifurcation diagram for solutions in a rectangular region is shown in Figure 4.6 (redrawn from Eilbeck, 1988). The ratio of the two sides of the rectangle is $\beta = \sqrt{2}$, and $d_1/d_2 = 0.3/0.033$. Although the diagram seems complicated, a short study reveals many features in common with Figures 4.1 and 4.3. The labelling of the main bifurcation curves follows a similar scheme: each curve is labelled (n,m) according to its main Fourier component near the bifurcation from the constant solution. In the rectangular region, the Fourier modes are $\cos \pi nx \cos \pi my/\beta$, so the label of (1,0) on the first branch corresponds to a solution which is constant in the y direction and proportional to $\cos \pi x$ in the x direction. Similarly the (0,1) branch, in the region near the constant solution, is proportional to $\cos \pi y/\beta$. The (1,1) branch is the first to show fully two-dimensional behaviour, behaving like $\cos \pi x \cos \pi y/\beta$ near the constant solution. The curves (1,0), (0,1) and (2,0) are identical in shape for the same reasons as in the one-dimensional case. The (1,1) curve will be copied as a (2,2) curve for larger values of α, but it has a different shape than the one-dimensional solution curves. However, the stability properties of the one-dimensional solutions in the two-dimensional region will be different from the case when they are constrained to one space dimension. For example, the (1,0) branch becomes unstable to perturbations in the y direction at $\alpha = 2.9713$ and evolves to one of the (0,1) branches. As in some of the 1-D cases, the choice of branch is arbitrary, due to simple symmetry considerations, and a particular choice can only be imposed by an external space-dependent parameter. The (0,1) branch for this particular choice of parameters becomes unstable at

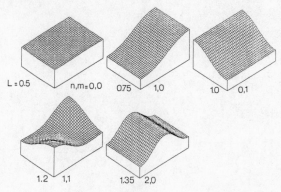

Figure 4.7.

$\alpha = 5.943$. Numerical studies for this case shows that the solution then evolves to the (2,0) branch.

This transition takes place on a rather slow time-scale, and may negate our assumption that such switches take place much faster than the rate of growth or the size of the object. With a sufficiently fast rate of growth, or a large perturbation, it may be possible to switch to the (1,1) branch instead. Figure 4.7 shows typical solution plots on the various branches, the rectangle in each case being scaled to the same size for graphical convenience.

As in the one-dimensional case, the arbitrariness in the switching from one branch to another may be resolved if cases with other values of d_1/d_2 are studied.

CONCLUSIONS

We have reviewed the well-known result that a succession of patterns can be generated as the size of the region containing the reaction–diffusion system grows. It is now straightforward to calculate a detailed chart of all the different patterns available to the growing system, and to determine which sequence will be followed in many cases. In some cases there is a degree of non-uniqueness as to the polarity of the sequence of patterns, but in other cases the polarity follows a fixed chain of events. It is clear that if at any time the system is given a large perturbation, the current stable pattern may be displaced and a new sequence followed. It would be interesting to examine the implications of this from a biological standpoint.

REFERENCES

Arcuri, P. and Murray, J.D. (1986) Pattern sensitivity to boundary and initial conditions in reaction–diffusion models. *J. Math. Biol. 24*, 141–165.
Brown, K.J. and Eilbeck, J.C. (1982). Bifurcation, stability diagrams, and

varying diffusion coefficients in reaction–diffusion equations.*Bull. Math. Biol.* *44*, 87–102.

De Dier, B. and Roose, D. (1987a) Determination of bifurcation points and catastrophies for the Brusselator model with two parameters. In T. Küpper *et al.* (eds.) *Bifurcation: Analysis, Algorithms, Applications* pp. 38–46. Basel: Birkhaüser.

De Dier, B., Walraven, R., Janssen, R., van Rompay, P. and Hlavacèk, V. (1987b) *Z. Naturforsch 42a*, 994–1004.

De Dier, B., Hlavacèk, V. and van Rompay, P. (1987c). Analysis of dissipative structures in a two-dimensional autocatalytic system: the Brusselator model. Preprint, Katholieke Universität Leuven.

Duncan, K. and Eilbeck, J.C. (1988) Numerical studies of symmetry-breaking bifurcations in reaction–diffusion systems. In L.M. Ricciardi (ed.) *Biomathematics and Related Computational Problems*, pp.439–48. Dordrecht: Kluwer.

Eilbeck, J.C. (1983) A Collocation approach to the numerical calculation of simple gradients in reaction–diffusion systems. *J. Math. Biol. 16*, 233–49.

—— (1987) Numerical studies of bifurcation in reaction–diffusion models using pseudo-spectral and path-following methods. In T. Küpper, R. Seydel and H. Troger (eds.) *Bifurcation: Analysis, Algorithms, Applications*, pp. 47–60. Basel: Birkhaüser.

—— (1988) Numerical studies of symmetry-breaking bifurcations in reaction–diffusion systems II. Heriot-Watt preprint.

Lane, D.C., Murray, J.D and Manoranjan, V.S. (1987) Analysis of wave phenomena in a morphogenetic mechanochemical model and an application to post-fertilization waves on eggs. *IMA J. Math. Appl. in Med. Biol. 4*, 309–31.

Meinhardt, H. (1982) *Models of Biological Pattern Formation*. London: Academic Press.

Ricciardi, L.M. (ed.) (1988) *Biomathematics and Related Computational Problems*. Dordrecht: Kluwer.

Sel'kov, E.E. (1968) Self-oscillations in glycolysis. *European J. Biochem. 4*, 79–86.

Turing, A.M. (1952). The chemical basis of morphogenesis. *Phil. Trans. R. Soc. B237*, 37–72.

5. Are there rules governing patterns of gene regulation?

MICHAEL A. SAVAGEAU

STRUCTURALISM AND HISTORICITY

Are there rules that govern the patterns observed among mechanisms of gene regulation? The answer depends upon whom one asks. There are some biologists who would answer: 'Of course there are rules, and it is the business of science to discover them!' This view has a long tradition embedded in positivist philosophy – collection of empirical data, induction of rules, synthesis of general laws. Brahe – Kepler – Newton provide the paradigm. This structuralist view was widespread in biology as well as physics a hundred years ago, and is still found today.

On the other hand, there are some biologists who would answer: 'No, there are no rules! Anything is possible. There is only what exists to be discovered and history'. This historical view is part of the Darwin legacy and, according to some, it has become the dominant view in modern biology. For a fuller account of these two philosophies in the context of developmental biology see Webster and Goodwin (1982).

In the realm of molecular genetics, the latter view has often been expressed explicitly by leaders in the field. One prominent pioneer in the study of gene regulation has stated that the rich variety of mechanisms governing gene expression is the result of historical accident. Nature is a tinker who haphazardly draws upon what already exists; she is not an engineer seeking optimal performance. Another well-known molecular geneticist has said that this rich variety shows that 'the only rule is that there are no rules'. Recently another authority has said, in addressing the question why are there positive and negative regulators, 'God only knows'.

In this article I would like to present one simple rule governing patterns of gene regulation that suggests how these two philosophies or views might be reconciled in a specific case. The rule deals with the question: why are there positive and negative modes of gene regulation? It turns out to be a very simple, logically deduced rule with strong predictive capabilities. It also has implications for nearly every area of biology. I shall describe briefly how this rule comes about, and then explore some of its implications for molecular mechanism, physiology, ecology, differentiation and evolution.

PATTERNS OF GENE REGULATION

Modern understanding of gene regulation begins with the work of Jacob and Monod (1961). Figure 5.1 summarizes their model of gene regulation in the lactose operon of *Escherichia coli*. The operon, or unit of transcription, consists of structural genes preceded by regulatory sites in the DNA. The co-ordinate expression of these genes is governed by the protein product of an associated regulatory gene. Note two features of this model for future reference: The regulator is a *negative* element in the system – a repressor – and its influence is exercised at the *initiation* of transcription. Many other examples have been shown to fit this model (Miller and Resnikoff, 1978). It has been extremely successful. Some might say that for a period it was too successful. The Jacob – Monod model completely dominated the study of gene regulation for nearly a decade. Everyone was looking for repressors, which made it difficult to recognize alternative designs.

Englesberg *et al* (1965) were the first to emphasize that the regulator in some systems is a positive element. This variation in design is shown in Figure 5.2. Nearly all other features of the classical Jacob – Monod model are the same – structural genes linked in a single transcriptional unit preceded by regulatory sites in the DNA. Regulation by an activator protein has been demonstrated for a number of other systems (Englesberg and Wilcox, 1974; Raibaud and Schwartz, 1984).

Studies with bacteriophage lambda showed that in some systems regulation is exerted at a termination site preceding the regulated structural genes (Roberts, 1969). The regulator in this case is the *N* gene product, which acts as an antiterminator or positive element in the regulation of

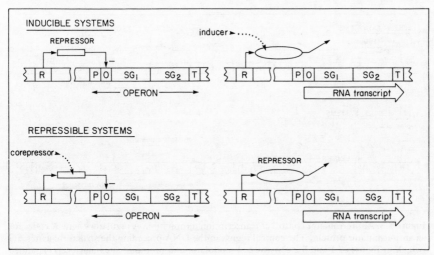

Figure 5.1. Repressor control of transcription initiation. A regulatory gene *R* codes for a repressor protein. Structural genes *SG* are preceded by a control region in the DNA consisting of a promoter site *P* and a modulator site *O*.

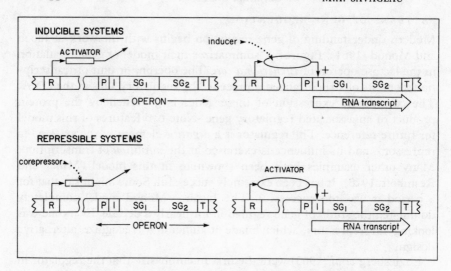

Figure 5.2. Activator control of transcription initiation. A regulatory gene *R* codes for an activator protein. Structural genes *SG* are preceded by a control region in the DNA consisting of a promoter site *P* and a modulator site *I*.

functions that are downstream from the terminator (see review by Friedman *et al.*, 1983). This type of mechanism is represented schematically in Figure 5.3. It also is common in amino acid biosynthetic systems where it is referred to as attenuation (Yanofsky, 1981).

To complete the symmetry, there is a third alternative in which the regulator acts as a proterminator or negative element to modulate transcription termination and readthrough into the adjacent structural genes. This

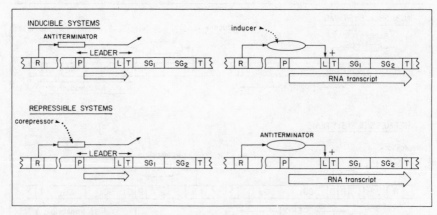

Figure 5.3. Antiterminator control of transcription termination. A regulatory gene *R* codes for an antiterminator protein. The control region in the DNA preceding the structural genes *SG* (which in Figures 5.1 and 5.2 consisted of modulator and promoter sites) has been expanded and redefined as the 'leader region', which begins with a promoter *P* and ends with a modulator site *L* and a terminator *T*. For simplicity, initiation of transcription is assumed to occur constitutively at the promoter.

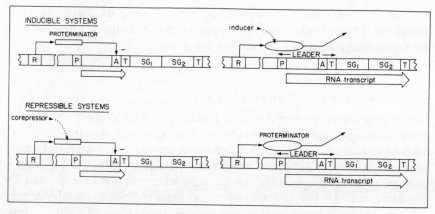

Figure 5.4. Proterminator control of transcription termination. A regulatory gene R codes for a proterminator protein. The leader region preceding the structural genes SG extends from the promoter P to the terminator T, with its associated modulator site A.

Table 5.1. Terminology

Mode of regulation	Regulator molecule	Modulator site	Transcript delimitor
Negative	Repressor	Operator	Promoter
Positive	Activator	Initiator	Promoter
Negative	Proterminator	Arrestor	Terminator
Positive	Antiterminator	Liberator	Terminator

variation was postulated earlier (Savageau, 1977), but until recently no examples had been reported. Perhaps the first is provided by the transcription-termination factor *rho*, which has been found to act as a proterminator in the negative regulation of its own structural gene (Brown et al., 1982). This type of mechanism is illustrated in Figure 5.4. The terminology for gene regulation in these figures is summarized in Table 5.1.

There are a number of other established variations in design for gene

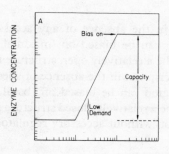
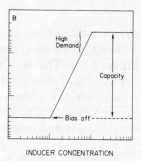

Figure 5.5. Steady-state induction characteristics of inducible catabolic operons. (a) Low demand regime for a system with a negative mode of regulation. (b) High demand regime for a system with a positive mode of regulation. (See text for basic concepts of gene regulation defined in terms of the induction characteristic.)

regulation, and rules are being deduced for some of these (e.g. see Savageau, 1979a; 1985c), but for the purposes of this article attention will be focused on differences in mode of regulation.

BASIC CONCEPTS OF GENE REGULATION

Before proceeding further it will be helpful to define a number of fundamental concepts that are used to characterize regulatory systems. This will be done in reference to Figure 5.5, which represents steady-state induction characteristics typical of inducible systems (Savageau, 1976, Ch. 14). Three distinct regions of operation can be seen when the data are plotted in log–log form. At low concentrations of inducer, below threshold, the system exhibits a basal level of expression independent of inducer concentration. At high concentrations of inducer, a maximal level of expression is produced independent of inducer concentration. At intermediate concentrations of inducer, the logarithm of expression is proportional to the logarithm of inducer concentration. This is the region of regulatable expression that is characterised by a relatively fixed amplification (logarithmic gain) between inducer (input) and gene expression (output).

Capacity

The ratio of the maximal to the basal level of expression represents the *capacity for regulation*. It can be large, 2 000-fold as in the case of the *lac* operon of *E. coli* (Sadler and Novick, 1965), or small, 18-fold as in the case of the *hut* operon of *Salmonella typhimurium* (Brill and Magasanik, 1969). This concept must be clearly distinguished from that of *demand for gene expression*, which is defined below.

Bias

This refers to the state of the system in the absence of any accessory regulatory elements. A promoter region can be biased on (maximal expression in Figure 5.5a). That is, it may be a relatively open structure that allows high levels of initiation and transcription in the absence of specific regulators. Alternatively, a promoter region can be biased off (basal expression in Figure 5.5b). That is, it may be a relatively closed structure that does not allow initiation and transcription without accessory regulators.

Mode

The mode of regulation may be either positive or negative; this duality of regulatory modes is fundamental to any regulatory system. To obtain regulated expression when the bias is off, a system must be supplied with an activator or positive regulatory element to increase expression above the basal level (Figure 5.5b). Expression then can be varied over the regulatable range by modulating the effectiveness of this accessory element. Alternatively, to obtain regulated expression when the bias is on, a system must be supplied with a repressor or negative regulatory element to decrease expression below the maximal level (Figure 5.5a). Expression can be varied from maximal to basal level in a regulatable fashion by modulating the effectiveness of the negative element. Thus, mode and bias are intimately related.

Demand

When a system operates at the high end of its regulatable range (Figure 5.5b) most of the time in its natural environment it is said to be a high-demand system. When it operates at the low end of its regulatable range (Figure 5.5a) most of the time in its natural environment it is said to be a low-demand system (Savageau, 1974; 1976; 1977). The concept of demand refers not only to a region of operation on the induction characteristic, but to the frequency with which it operates in that region. It is thus analogous to the concept of 'duty cycle' used in other fields to refer to the fraction of time that a device is in operation.

WHY POSITIVE AND NEGATIVE MODES OF GENE REGULATION?

Is it historical accident, as many have claimed? After all evolution is opportunistic. An organism happens upon a particular mechanism at some point, it works as well as any other, and so it is maintained. Alternatively, do these designs imply functional differences that lead to their selection in specific contexts? If so, can one discern what these functional differences might be and make predictions about the conditions that lead to their selection?

These types of biological questions are representative of an important class that are difficult to answer using *only* the direct experimental approach (Savageau, 1976; 1977). One cannot simply do the same experiments on two typical systems – say the maltose (positive) and the galactose (negative) operons – compare the results and draw meaningful conclusions. As any good experimentalist knows, this is not a well-controlled experiment. These systems differ in a variety of ways that have nothing to do with the question of positive v. negative regulation. These systems code for a different number of enzymes, the enzymes catalyse different types of reactions, they have different kinetic properties, etc. These factors would tend to obscure the interpretation of any differences observed and make it difficult to attribute differences to the mode of regulation *per se*.

Ideally one would like a well-controlled comparison in which the two systems were identical in every respect, except that one has a positive and the other a negative mechanism of regulation. While one might conceive of constructing the alternatives by appropriate genetic engineering, this is not easily accomplished. At the present time it is much more practical to simulate such comparisons by appropriate mathematical analysis of gene regulation. One formulates models with positive and negative mechanisms of gene regulation, makes all other features of the models identical by fiat, and exhaustively analyses the models to determine the irreducible differences. Thus, this approach represents an ideal controlled comparison (Savageau, 1972; 1976; 1985b; Irvine and Savageau, 1985).

The power-law formalism, which provides the basis for the comparative approach referred to above, has an extensive literature beginning in the late 1960s (e.g. Savageau, 1969; 1971; 1972; 1976; 1979b; 1985a; Voit and Savageau, 1987; Sorribas and Savageau, 1988) and is beyond the scope of this article. The interested reader may wish to begin with a current review dealing with recent developments on the subject (Savageau and Voit, 1987).

DEMAND THEORY

The rule that summarises the demand theory of gene regulation may be stated as follows. The molecular mode of gene regulation is correlated with the demand for gene expression in the organism's natural environment – positive when the demand is high and negative when the demand is low. The development of this theory involves elucidation of the functional differences between positive and negative mechanisms, determination of the mutational tendencies within these mechanisms, assessment of the population dynamics in mixed cultures of wild-type and mutant organisms, and prediction of conditions that lead to the selection and maintenance of functional regulatory mechanisms. These stages are outlined below.

Functional Differences

Detailed analysis of mechanisms with either a positive or a negative mode of regulation shows that in most respects they behave identically; the inherent differences in function appear in response to mutations in the regulatory mechanism itself (Savageau, 1974; 1976). If there is to be a functional explanation for the selection of positive and negative modes of gene regulation, then this analysis suggests that one must focus on the behaviour of mutants altered in the regulatory mechanism.

Mutational Entropy

Most mutations in a highly evolved structure are detrimental; few are beneficial. This well-known observation from classical genetics is simply another manifestation of the entropy principle (Savageau, 1974; 1976; 1977). Consequently, most mutations in a regulatory mechanism lead to loss of regulation. In the case of a positive mode, loss of regulation results in super-repressed expression because the built-in bias of the promoter region is off. In the case of a negative mode, loss of regulation results in constitutive expression because the built-in bias is on. Thus, mechanisms with alternative modes of regulation will respond in diametrically opposed ways to the predominant mutational tendencies that influence them.

Population Dynamics

The question of how wild-type and mutant organisms will fare in a common environment is answered by examining their population dynamics under different environmental conditions (Savageau, 1974; 1976; 1977; 1980). The results are summarised in Table 5.2. There are four panels to consider.

Consider a mechanism with a positive mode in a high-demand environment; for example, the maltose catabolic operon of *E. coli* (Hofnung and Schwartz, 1971) in an environment where maltose is the sole carbon source. The wild-type organism will induce the maltose catabolic enzymes to a high level and grow at a rapid rate. The mutants, which have lost regulation of this system, will be unable to grow because the maltose promoter is biased off. As a result the mutants will be culled from the population and the functional regulatory mechanism will be selected.

In a low-demand environment (e.g. when succinate is the sole carbon source) expression of the *maltose* operon will be off in the wild-type because there is no physiological inducer. The mutants, which have lost regulation of this system, will be unable to express the maltose catabolic enzymes because the maltose promoter is biased off. There is no selection of regulatory mechanisms for the functional regulatory mechanism under these conditions. Super-repressed mutants will accumulate with time as a result of mutational entropy, and the functional regulatory mechanism will be lost through genetic drift.

The results are just the opposite when one considers a mechanism with

Table 5.2. Predicted correlation between molecular mode of regulation and demand for gene expression

Demand for expression	Mode of regulation	
	Positive	Negative
High	Regulation selected	Regulation lost
Low	Regulation lost	Regulation selected

Table 5.3. Physiology and demand for gene expression

Type of system	Condition corresponding to high demand
Repressible scavenging pathway	Substrate seldom present in high concentrations
Repressible drug sensitivity	Drug frequently present in high concentrations
Repressible biosynthetic pathway	End product seldom present in high concentrations
Inducible biosynthetic enzyme	End product seldom present in high concentrations
Inducible catabolic pathway	Substrate frequently present in high concentrations
Inducible detoxification pathway	Substrate frequently present in high concentrations
Inducible genetic exchange	Genetic exchange frequently occurs
Inducible repair response	Repair response frequently required
Inducible toxin production	Toxin frequently produced
Inducible biosynthesis of surface antigen	Antigenic determinants frequently required

a negative mode; for example, the galactose catabolic operon of *E. coli* (Nakanishi *et al.*, 1973). In a high-demand environment with galactose as the sole carbon source, expression of the galactose operon will be induced to high levels in the wild-type organism. Expression in the mutants also will be at high levels because the bias of the galactose promoter is on. There is no selection for the functional regulatory mechanism under these conditions. Constitutive mutants will accumulate with time as a result of mutational entropy, and the functional regulatory mechanism will be lost through genetic drift.

In a low-demand environment (e.g., when succinate is the sole carbon source) expression of the galactose operon will be off in the wild-type because there is no physiological inducer. The mutants, which have lost regulation of this system, will be expressing the galactose catabolic enzymes at high levels because the galactose promoter is biased on. Such high-level expression at inappropriate times or locations leads to dysfunction. Energy and material are wasted in synthesizing unnecessary proteins, and inappropriate high-level expression interferes with normal functions in the cell. As a result the mutants will grow slower. They will be culled from the population and organisms with the functional regulatory mechanism will be selected.

Predictions of demand theory

The diagonal (upper left and lower right entries) in Table 5.2 yields the predicted correlation of positive mode with high demand for expression in the organism's natural environment and negative mode with low demand. Only the regulatory mechanism in active use overcoming the built-in bias of the system will exhibit physiological dysfunction when mutated and thus be subject to natural selection.

The off-diagonal entries (lower left and upper right) in Table 5.2 yield the predicted loss of a functional regulatory mechanism through genetic drift and loss of regulation when the demand for expression dictated by the environment is inappropriate for the mode of regulation. If the regulatory mechanism is not in active use overcoming the built-in bias of the system it will not exhibit dysfunction when mutated and thus not be subject to natural selection.

The theory outlined above is stated in general terms and consequently embodies a certain level of abstraction. In order to make the implications of this theory explicit one must know what high and low demand mean in different physiological contexts.

PHYSIOLOGY

Table 5.3 lists a variety of physiological functions and the corresponding conditions for high demand. With a little thought one can generally determine the conditions that correspond to high demand for any particular physiological function of interest.

For example, in the above presentation of demand theory, inducible catabolic systems were considered; for these systems, high demand corresponds to substrate frequently present at high concentrations. However, if one considers repressible biosynthetic systems, the physiology is different and the condition corresponding to high demand for expression must be interpreted accordingly. When an amino acid is absent from an organism's natural environment, the organism must synthesize that amino acid endogenously. Expression of the corresponding amino acid biosynthetic operon is in high demand. Thus, high demand in this physiological context corresponds to end product seldom present at high concentrations in the organism's natural environment. Other examples are described in Savageau (1983a).

ECOLOGY

In order to relate physiology to demand it is necessary to have information about the organism's natural environment. Since most of the systems that are well studied at the molecular level are found in enteric bacteria and their phages, one can focus on the environment of these organisms and produce a large number of results with which to test the implications of demand theory.

One of the principal habitats of *E. coli* is the colon of warm-blooded animals. This is a rather complex ecosystem (Drasar and Hill, 1974; Freter, 1976) determined by the diet, physiology and immunological state of the host. It also is determined by abundant interactions among the other 500 or so micro-organisms that are found in the colon and by complex physical and geometrical factors. Nevertheless, one can estimate the relative abundance of various nutrients in the colon by two types of measurements, one an indirect method and the other direct (Savageau, 1974; 1977; 1983c).

Relative abundance of nutrients

If one starts with nutrients present in the diet and measures the extent to which they are removed by the host before the intestinal contents reach the colon, one can determine indirectly the relative abundance of nutrients to which the bacteria are exposed. For example, dietary sugars like galactose are efficiently absorbed early in the small intestine and consequently very little will survive transit to the colon and be made available to the bacteria. Inducible systems in the bacteria that allow catabolism of these sugars will be in low demand. Other sugars like arabinose are poorly absorbed by the host and will reach the colon with little attenuation in concentration. Inducible systems in the bacteria for these sugars will tend to be in high demand.

Other things being equal, the greater the affinity of the host-transport system in the small intestine, the lower the relative concentration of the corresponding sugar in the colon. From such indirect measurements one can rank the relative abundance of various sugars in the colon as follows (Savageau, 1974): D-glucose < D-galactose < glycerol < D-xylose < L-glycose < L-mannose < L-fucose < L-rhamnose < L-arabinose. For additional indirect estimates of nutrient abundance see Savageau (1976, 1977, 1983c).

More direct estimates of relative abundance have been made by using a catheter to remove intestinal contents for assay (Nixon and Mawer, 1970). By this method the relative abundance of amino acids in the colon is determined to be: lysine > glutamate > arginine > tyrosine > trypto-phan > glycine > leucine > phenylalanine > histidine > alanine > serine > valine > aspartate > proline > threonine > cystine > isoleucine > methionine (see also Savageau, 1983c).

TESTS OF DEMAND THEORY

With information at hand concerning the molecular mechanisms of gene regulation and the relative abundance of nutrients in the microenvironment of enteric bacteria, one can test the predicted correlation of regulatory mode at the molecular level and the demand for gene expression in the natural environment.

Results for inducible catabolic systems are summarized in Table 5.4. For example, galactose is seldom present at high concentrations in the colon, which corresponds to low demand for the galactose catabolic operon inducible system in *E. coli*. According to demand theory, the galactose system should have a negative mode of regulation (Savageau, 1974), which is in agreement with the classical repressor that has been found to govern this system (Nakanishi *et al.*, 1973). On the other hand, arabinose is frequently present at high concentrations in the colon, which corresponds to high demand for the arabinose catabolic operon inducible system in *E. coli*. Thus, one predicts a positive mode of regulation for this system (Savageau, 1974), which is in agreement with the finding that the major regulation of this system occurs by an activator protein (Englesberg and Wilcox, 1974).

Similar results are shown in Table 5.5 for a number of repressible biosynthetic systems. Tryptophan biosynthetic operon is an example involving an amino acid that is relatively abundant in the colon. The corresponding biosynthetic operon in the bacterium is in low demand, and demand theory allows one to predict a negative mode of regulation (Savageau, 1976; 1977). This is in agreement with molecular studies showing that the predominant mechanism of regulation for the tryptophan operon in *E. coli* is a classical repressor (Squires *et al.*, 1975). On the other hand, the histidine biosynthetic operon is typical of those involving amino acids that are seldom present at high concentrations in the colon. The corresponding biosynthetic system in the bacterium is in high demand, and one predicts a positive mode of regulation (Savageau, 1976; 1977). Again the prediction is in agreement with molecular data, in this instance indicating regulation by an antiterminator or attenuator mechanism (DiNocera *et al.*, 1978; Barnes, 1978; Johnston *et al.*, 1980).

Table 5.4. Molecular mode of regulation correlates with the demand for gene expression in the organism's natural environment*

Inducible catabolic systems	Mode of regulation	Demand for expression
Arabinose	Positive	High
Deoxyribonucleotides	Negative	Low
Galactose	Negative	Low
Glycerol	Negative	Low
Histidine	Negative	Low
Lactose	Negative	Low
Maltose	Positive	High
Rhamnose	Positive	High
Xylose	Positive	High

* Enteric bacteria in the mammalian colon

Table 5.5. Molecular mode of regulation correlates with the demand for gene expression in the organism's natural environment[*]

Repressible biosynthetic systems	Mode of regulation	Demand for expression
Arginine	Negative	Low
Cysteine	Positive	High
Histidine	Positive	High
Isoleucine-valine	Positive	High
Leucine	Positive	High
Lysine	Negative	Low
Phenylalanine	Positive	High
Threonine	Positive	High
Tryptophan	Negative	Low
Biotin	Negative	Low

[*] Enteric bacteria in the mammalian colon

The results of additional tests that have been reported elsewhere (Savageau, 1979a; 1983a) are summarized in Table 5.6. From all these results one can conclude that the predicted correlation between molecular mode of regulation and environmental demand for gene expression is very strong. There are no well-documented exceptions that involve good data at both the molecular level concerning the mode of regulation and the ecological level concerning the demand for gene expression. The alternative hypothesis – that mode of regulation is the result of historical accident – is very unlikely. The probability is 2^{-n} that the predictions of demand theory would fortuitously agree in n cases if the mode of regulation were randomly determined by historical accident. This is approximately 1 in 10^6 for an n of 20.

If, on the basis of the evidence cited above, one tentatively accepts demand theory as the best explanation currently available for positive and

Table 5.6. Summary of experimental data for molecular mode of regulation, physiology, ecology, and demand for gene expression[*]

54	Gene systems examined
10	Types of physiology represented
32	Systems known or predicted to be in low demand in the organism's natural environment
22	Systems known or predicted to be in high demand in the organism's natural environment
40	Systems with known modality and demand that are in agreement with the predictions of demand theory
9	Systems with known modality but unknown demand for which demand theory yields testable predictions
5	Systems with known demand but unknown modality for which demand theory yields testable predictions

[*] Savageau (1983c)

negative modes of gene regulation, then a number of additional implications follow.

COUPLED MOLECULAR MECHANISMS

The demand for expression of certain genes in the cell is logically coupled to that of other genes. The mode of regulation of these logically coupled genes has a predictable pattern according to demand theory (Savageau, 1985c). Knowledge of the mode of regulation for one gene allows prediction of the mode of regulation for other logically coupled genes. The examples in Table 5.7 will make this clear.

The high-capacity transport system bringing the sugar arabinose into the cell (Kolodrubetz and Schleif, 1981) and the arabinose catabolic operon for the utilization of intracellular arabinose (Englesberg and Wilcox, 1974) are functions for which the demand is directly coupled. It would be dysfunctional to induce one without the other. According to demand theory, expression of these logically coupled gene systems should be governed by the same mode of regulation.

The low-capacity scavenging system for bringing the amino acid tyrosine into the cell (Whipp and Pittard, 1977) functions in parallel with the tyrosine biosynthetic operon (Camakaris and Pittard, 1982). When tyrosine is abundant in the environment of the cell there is little need for either scavenging or endogenous biosynthesis, and the demand for expression of these gene systems will be accordingly low.

The isoleucine–valine biosynthetic system provides an example of a repressible biosynthetic pathway (Nargang *et al.*, 1980; Lawther and

Table 5.7. Mode of gene regulation for logically coupled functions and the associated demand for their expression*

Functions	Genes	Mode	Demand
Transport	*araE*	Positive	High
Catabolism	*araB*	Positive	High
Scavenging	*tyrP*	Negative	Low
Biosynthesis	*tyrA*	Negative	Low
Inducible biosynthesis	*ilvC*	Positive	High
Repressible biosynthesis	*ilvA*	Positive	High
Initial stage of catabolism	*uxaC*	Negative	Low
Final stage of catabolism	*kdgA*	Negative	Low
Biosynthesis	*hisG*	Positive	High
Catabolism	*hutH*	Negative	Low

*Enteric bacteria in the mammalian colon

Hatfield, 1980) in the middle of which there is an enzyme that is substrate-induced (Biel and Umbarger, 1981). The logic of this arrangement assures that a high demand for expression of the biosynthetic pathway is accompanied by a high demand for expression of the inducible enzyme.

In converging catabolic pathways, the demand for the final stages is logically related to the demand for the initial stages. A low demand for the final stages cannot be present with a high demand for one of the initial stages. The low demand for the *kdg* genes, which code for the final enzymes in the catabolism of both galacturonate and gluconate (Pouyssegur and Stoeber, 1974), implies a low demand for the *uxa* genes, which code for the initial enzymes in the catabolism of galacturonate (Portalier *et al.*, 1980).

In the examples given so far, the demand for the logically coupled genes was the same. By contrast, there is a reciprocal demand for biosynthetic and catabolic systems that share a common metabolite as product and substrate, respectively. For example, the histidine biosynthetic enzymes (DiNocera *et al.*, 1978; Barnes, 1978; Johnston *et al.*, 1980) are in high demand while the histidine utilization enzymes (Brill and Magasanik, 1969; Smith and Magasanik, 1971) are in low demand in the absence of exogenous histidine. Physiological dysfunction would result if these two systems were to be simultaneously off or on.

In each of these examples the experimentally-determined modes of regulation agree with those predicted by demand theory. In each case, one readily can deduce the logical coupling and thus the pattern of demand for the logically coupled genes. Even in cases where the individual demands are unknown, the *pattern* still can be deduced. In such cases, experimental determination of the molecular mode of regulation for one gene allows immediate prediction of the mode for the other logically coupled genes.

DIFFERENTIATION

Another important class of logically coupled molecular mechanisms is differentiated cell-specific gene functions. The paradigm is shown in Table 5.8. A-specific functions in A cells tend to be frequently expressed at high levels; they define what it means to be an A cell. Consequently, they are in high demand and demand theory predicts a positive mode of regulation.

Table 5.8. Mode of regulation predicted by demand theory for cell-specific genes in differentiated cell types

Cell type	Cell specific genes	
	A	B
A	Positive	Negative
B	Negative	Positive

B-specific functions in A cells tend to be seldom expressed at high levels; by definition they are off in A cells. Their expression is in low demand and a negative mode of regulation is predicted for these functions in A cells. Similar results are predicted for the modes of regulation of cell specific functions in B cells. This is a straightforward application of demand theory.

In addition to predicting the mode for each regulatory mechanism, demand theory predicts a *switching* of the regulatory mechanisms themselves during differentiation (Savageau, 1980; 1983b). When A cells differentiate into B cells, A-specific functions are turned off and B-specific functions are turned on. If one considers a given set of cell-specific functions, say the A-specific functions in Table 5.8, one sees the demand regime change and the theory predicts a *switching* from regulatory mechanisms with one modality to those with the opposite modality.

Evidence in support of these predictions comes from model systems that are well studied at the molecular level in prokaryotes, phages, and lower eukaryotes. Two examples will be considered here: trytophan biosynthesis in colon-type and aquatic-type *E. coli* and viral gene expression in lysogenic-type and lytic-type lambda phage.

Escherichia coli

E. coli is not normally considered to exhibit differentiated cell types. However, this organism has at least two distinct habitats that differ in their physical and chemical character, including marked differences in the spectrum and level of nutrients (Savageau, 1983c). *E. coli* cells found in the colon express different sets of genes than *E. coli* cells found in lakes and streams, and in this sense may be considered 'differentiated' cell types (Savageau, 1983b). When an aquatic-type *E. coli* cell colonizes the colon of an animal it 'differentiates' into a colon-type *E. coli* by turning off certain genes and turning on others.

For example, the colon habitat is relatively rich in tryptophan, whereas the aquatic habitat provides relatively little free tryptophan. The tryptophan biosynthetic genes will be expressed in aquatic-type *E. coli* because this cell type must make all its own tryptophan; these same genes will not be expressed in colon-type *E. coli* because this cell type receives free tryptophan from its environment. Thus, expression of the tryptophan biosynthetic genes may be considered a differentiated cell-specific function associated with aquatic-type *E. coli* (Savageau, 1983c).

According to demand theory, the tryptophan biosynthetic operon is expected to have a negative mode of regulation in colon-type cells (low demand), to have a positive mode of regulation in aquatic-type cells (high demand), and to undergo switching of regulatory mechanisms during differentiation (Savageau, 1983b). These expectations are in agreement with experimental data showing that the tryptophan biosynthetic system is governed by a classical repressor mechanism (negative mode) when the amino acid is abundant (Squires *et al.*, 1975) and by an antiterminator mechanism (positive mode) when it is scarce (Yanofsky, 1981) (Figure 5.6). This application of demand theory is important not so much for what it reveals about the differentiation process in higher organisms, but for the rationale it provides for dual regulatory mechanisms of opposite modality in bacteria.

Bacteriophage lambda

A more familiar model system for the study of differentiation is provided by bacteriophage lambda, a temperate virus that exhibits two different life styles (Herskowitz and Hagen, 1980). When growing as a lysogen, the phage is stably integrated into the host chromosome and only a few lysogen-specific viral genes are expressed. When the phage is induced to grow lytically, lysogen-specific genes are turned off and lytic-specific genes are turned on in a programmed fashion.

According to demand theory, one expects lysogen-specific genes (high demand) to have a positive mode of regulation and lytic-specific genes (low demand) to have a negative mode of regulation in the lysogen; lysogen-specific genes (low demand) to have a negative mode of regulation and

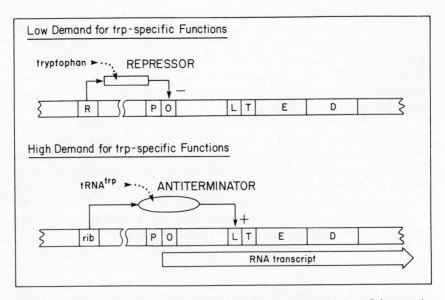

Figure 5.6. Switching of a regulatory mechanism with one modality to that of the opposite modality during differentiation. The tryptophan operon of *E. coli* in colon- and aquatic-type cells. (See text for discussion.)

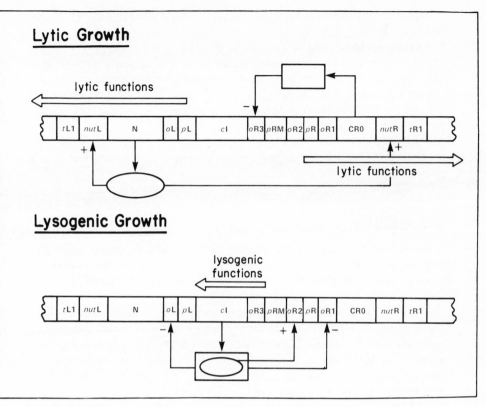

Figure 5.7. Switching of regulatory mechanisms with one modality to those of the opposite modality during differentiation. Regulation of viral gene expression in bacteriophage lambda during lytic and lysogenic growth.

lytic-specific genes (high demand) to have a positive mode of regulation in the lytic virus, and switching of regulatory mechanisms during differentiation (Savageau, 1983b).

The regulatory mechanisms in lambda are well characterized at the molecular level and are in agreement with the expectations from demand theory (Figure 5.7). The lytic genes are not expressed in the lysogen; they are in low demand and subject to regulation by the CI repressor (negative mode) (Ptashne *et al.*, 1980). During lytic growth, the lytic genes are expressed; they are in high demand and subject to regulation by an antiterminator, the *N* gene product (positive mode) (Friedman *et al.*, 1983). Similarly, the product of the lysogen-specific gene *CI*, which in the lysogen is in high demand, is an activator for transcription of its own structural gene (positive mode) (Ptashne *et al.*, 1980). During lytic growth, the *CI* gene is not expressed; it is in low demand and governed by CRO repressor (negative mode) (Ptashne *et al.*, 1980). The switching from regulatory mechanisms of one modality to those of the opposite modality during differentiation is brought about by the interlocking circuitry of gene regulation in lambda.

As these examples and others (Savageau, 1983b) show, demand theory is able to account for the modes of regulation of differentiated cell-specific functions and the switching of regulatory mechanisms during differentiation in a number of well-studied model systems. The evidence in the case of higher organisms is too preliminary to draw any conclusions at this point. Nevertheless, the frequent observation of alternative promoters upstream of eukaryotic structural genes (Schibler and Sierra, 1987) suggests that dual regulatory mechanisms of opposite modality might be a common theme in all organisms.

EVOLUTION

A useful model for the experimental study of evolution is the acquisition of new metabolic functions in bacteria. This manifestation of evolution is termed *acquisitive evolution*. The paradigm is evolved β-galactosidase in *E. coli*.

If one deletes the entire lactose operon from the *E. coli* genome and plates the mutant cells on medium containing lactose as the sole source of carbon, one finds that the cells no longer grow as the wild-type did. However, if one waits awhile and looks carefully one finds that small colonies arise at a low frequency. How do the mutant cells acquire the ability to catabolize lactose? The answer is they appropriate some other existing function in the cell and remodel it for hydrolysis of lactose.

Acquisition of activity

Since the mutant is initially unable to grow on lactose, it is clear that no function actively expressed and used in the cell has the ability to hydrolyse lactose at a rate sufficient to support cellular growth. The function to be acquired must be increased in activity. Increasing the activity of a function actively expressed and used in the cell will tend to be counter-productive because it will compromise the integrity of the original function (Wright, 1931). This is well-known from experience in animal and plant breeding. Attempts to select simultaneously for two or more traits tend to be very inefficient. Hence, it is much more likely that the ability to utilize lactose will be acquired from an existing function that is not in active use.

Not in active use implies low demand for expression, and according to demand theory such a function will have a negative mode of regulation. Mutational entropy implies that constitutive mutants will be continuously thrown off at a reasonable frequency. Normally, such mutants would be selected against and the functional negative mode of regulation maintained when there is a low demand for expression in the organism's natural environment. However, with the normal lactose operon deleted, there is now a high demand for expression of the newly acquired function, and constitutive mutants in this system are strongly selected.

Loss of the negative mode of regulation for a function that is normally not expressed leads to an instantaneous amplification of that activity – by as much as 2 000-fold in some cases. This amplification implies that significant lactose hydrolysis can be obtained from an enzyme with only a weak non-specific (or parasitic) activity for lactose. In the case of evolved β-galactosidase this is sufficient to support the slow growth of cells on lactose.

As indicated above, the initial remodeling of the acquired function can occur by a single mutation leading to the loss of regulation. Once the cell has survived the initial crisis and is able to grow on lactose, remodelling of the acquired function can occur by mutations that enhance catalytic activity. For example, mutations in the structural gene for the enzyme can enhance its affinity and molecular activity for the new substrate lactose. This may require multiple mutations that occur in a step-wise fashion when the original activity is only distantly related to β-galactosidase. Gene duplication might also contribute to the initial amplification of gene expression in activity, but its more important role is to allow for divergent evolution of one gene copy while maintaining one copy of the original function.

Acquisition of regulation

Once the cell has acquired substantial activity for growth on lactose, selection will favour the evolution of regulatory mechanisms that further enhance performance of the appropriated function. According to demand theory, most newly evolved functions will have the remnant of a regulatory mechanism with a negative mode. Given that the recent history of the function involved a low demand regime, the most readily acquired mechanism of regulation will be a remodelled version of the previously existing mechanism with a negative mode. For this to occur it is necessary to revert the original mutation leading to constitutive expression. It also is necessary to alter the regulator so that it will recognize lactose or some derivative as an inducer. Again, this is likely to require multiple mutations if the original regulator has little affinity for lactose.

If there were a low demand for lactose hydrolysis in the environment of such an organism, as is true of the natural environment of $E.$ $coli$, then there would be selective pressure to maintain a functional regulatory mechanism with a negative mode. However, if in the long-term environment there were a high demand for lactose hydrolysis, a functional regulatory mechanism with a negative mode could not be maintained. There would be selective pressure favouring the evolution and then maintenance of a regulatory mechanism with a positive mode.

The evolution of a regulatory mechanism with a positive mode from one with a negative mode is likely to be a slow process. Remodelling of regulatory design will be required not only of the regulator but also of the regulatory region preceding the regulated structural genes. For example, a repressor must be changed to an activator, and a promoter with a bias on must be changed to one with a bias off. These are substantial changes that will require multiple mutations in an appropriate sequence. Alternatively, it is possible for this to happen as follows. If the regulator and corresponding regulatory sequence of a functioning activator-controlled system are considered as a unit, then this unit could be duplicated, transposed and fused to the newly acquired structural genes. In this process, high levels of expression could be maintained during the conversion of a non-functional regulatory mechanism with a negative mode (constitutive expression) to a non-functional regulatory mechanism with a positive mode (expression turned on by an inappropriate inducer). Following this single transpositional event, the non-functional positive mode of regulation could be made functional by mutations in the positive regulator that increase its affinity for lactose while decreasing its affinity for the original inducer. In a high-demand environment there would be selective pressure favouring such processes and maintaining the resultant regulatory mechanism with a positive mode.

There is evidence, from the experimental study of acquisitive evolution in bacteria, for each of the steps described above (Novick, 1958; Novick and Weiner, 1959; Horiuchi *et al.*, 1963; Rigby *et al.*, 1974; Clarke, 1978; Mortlock, 1982; Hall, 1982), except for the conversion of a regulatory mechanism with a negative mode of regulation conversion to one with a positive mode. The reason for this exception is undoubtedly that no one has looked for such an event. Demand theory provides a rationale for designing experiments that can be expected to yield examples of conversion in modality of regulation.

CONCLUSION

In this article a simple rule governing the mode of gene regulation has been described. The positive mode is associated with genes whose expression is in high demand in the natural environment, whereas the negative mode is associated with genes whose expression is in low demand. This rule is the basis for the demand theory of gene regulation. The development of the theory has been outlined, and the implications have been traced through several areas of biology. The theory is seen to provide readily testable predictions in all of these areas and to integrate and systematize a large number of otherwise disparate observations.

In some cases the mode of regulation for a recently acquired function may not agree with the new demand for its expression. Such discrepancies reflect historical contingencies associated with the origins of the mechanism. While such discrepancies may be evident initially, they cannot be expected to survive the long-term selective pressures that enforce the rule of demand theory.

Differences were also seen in the detailed molecular mechanisms by which a given mode is realized and by which the switching of modality is brought about in particular cases. Such differences might be the result of historical accidents that are functionally neutral, or they might be governed by additional rules that have yet to be determined. One can always assume that certain differences are the result of historical accident, but such an explanation has no predictive power and tends to stifle the search for alternative hypotheses. It generally tends to be more productive if one starts with the working hypothesis that there are rules. One may end up attributing differences to historical accident, but in my opinion it is a mistake to start there.

Accident and rule both play a role, but generally at different levels and during different periods. The result is plasticity in biological organization, but within bounds. Nature is optimizing, but subject to constraints (as is true of all optimization when properly understood). Mutation creates the diversity over which optimization acts; selection is the optimiser. Both the processes of mutation and selection are subject to constraints. External constraints are provided by origins and historical contingencies, while internal constraints are imposed by the laws of physics and the logical and dynamical laws of system organization.

ACKNOWLEDGEMENTS

This article is an outgrowth of a sabbatical year spent in Canberra, Australia. I thank Drs I. Young and P. Benyon for making my visit possible, for numerous stimulating discussions, and for their generous hospitality. I also thank Drs S. Cooper and M.H. Meisler for helpful suggestions during the preparation of the manuscript. The work was supported in part by funds from CSIRO in Australia and grants from the Presidential Initiatives Fund of the University of Michigan, the National Institutes of Health, and the National Science Foundation in the US.

REFERENCES

Barnes, W.M. (1978) DNA sequence from the histidine operon control region: seven histidine codons in a row. *Proc. Natl. Acad. Sci. 75*, 4281–5.

Biel, A., and Umbarger, H.E. (1981) Mutations in the *ilvY* gene of *Escherichia coli* K-12 that cause constitutive expression of *ilvC*. *J. bacteriol. 146*, 718–24.

Brill, W.J., and Magasanik, B. (1969) Genetic and metabolic control of histidase and urocanase in *Salmonella typhimurium*, strain 15–59. *J. Biol. Chem. 244*, 5392–402.

Brown, S., Albrechtsen, B., Pedersen, S., and Klemm, P. (1982) Localization and regulation of the structural gene for transcription-termination factor rho of *Escherichia coli.*, *J. Mol. Biol. 162*, 283–98.

Camakaris, H., and Pittard, J. (1982) Autoregulation of the *tyrR* gene. *J. Bacteriol. 150*, 70–5.

Clarke, P.H. (1978) Experiments in microbial evolution. *The Bacteria 6*, 137–218.

DiNocera, P.P., Blasi, F., DiLauro, R., Grunzio, R., and Bruni, C.B. (1978) Nucleotide sequence of the attenuator region of the histidine operon of *Escherichia coli* K-12. *Proc. Natl. Acad. Sci. 75*, 4276–80.

Drasar, B.S., and Hill, M.J. (1974) *Human Intestinal Flora*, New York: Academic Press.

Englesberg, E., and Wilcox, G. (1974) Regulation: positive control. *Annu. Rev. Genet. 8*, 219–42.

Englesberg, E., Irr, J., Power, J., and Lee, N. (1965) Positive control of enzyme synthesis by gene *C* in the L-arabinose system. *J. Bacteriol. 90*, 946–57.

Freter, R. (1976) Factors controlling the composition of the intestinal microflora. In H.M. Stiles, W.J. Loesche and T.C. O'Brien, (eds.) *Proceedings of the Microbial Aspects of Dental Caries*, special supplement, *Microbiology Abstracts*, vol. 1, pp.109–120.

Friedman, D.I., Schauer, A.T., Mashni, E.J., Olson, E.R., and Baumann, M.F. (1983) *Escherichia coli* factors involved in the action of the λ gene *N* antitermination function. In D. Schlessinger (ed.) *Microbiology 1983*, pp.39–42. Washington, D.C.: American Society for Microbiology.

Hall, B. (1982) Evolution on a petri dish: the evolved beta-galactosidase system as a model for studying acquisitive evolution in the laboratory. *Evol. Biol. 15*, 85–150.

Herskowitz, I., and Hagen, D. (1980) The lysis–lysogeny decision of phage λ: explicit programming and responsiveness. *Annu. Rev. Genet. 14*, 399–445.

Hofnung, M., and Schwartz, M. (1971) Mutations allowing growth on maltose of *Escherichia coli* K12 strains with a deleted *malT* gene. *Mol. Gen. Genet. 112*, 117–132.

Horiuchi, T., Horiuchi, S., and Novick, A. (1963) The genetic basis of hypersynthesis of β-galactosidase. *Genetics 48*, 157–69.

Irvine, D.H., and Savageau, M.A. (1985) Network regulation of the immune response: alternative control points for suppressor modulation of effector lymphocytes. *J. Immun. 134*, 2100–2116.

Jacob, F., and Monod, J. (1961) Genetic regulatory mechanisms in the synthesis of proteins. *J. Mol. Biol. 3*, 318–56.

Johnston, H.M., Barnes, W.M., Chumley, F.G., Bossi, L., and Roth, J.R. (1980) Model for regulation of the histidine operon of *Salmonella*. *Proc. Natl. Acad. Sci. 77*, 508–12.

Kolodrubetz, D., and Schleif, R. (1981) Regulation of the L-arabinose transport operons in *Escherichia coli*. *J. Mol. Biol. 151*, 215–27.

Lawther, R.P., and Hatfield, W.W. (1980) Multivalent translational control of transcription termination at attenuator of *ilvGEDA* operon of *Escherichia coli* K-12. *Proc. Natl. Acad. Sci. 77*, 1862–6.

Miller, J.H., and Reznikoff, W.S. (1978) *The Operon*. Cold Spring Harbor, N.Y.: Cold Spring Harbor Laboratory.

Nakanishi, S., Adhya, S., Gottesman, M.E., and Pastan, I. (1973) *In vitro* repression of the transcription of *gal* operon by purified *gal* repressor. *Proc. Natl. Acad. Sci. 70*, 334–8.

Nargang, F.E., Subrahmanyam, C.S., and Umbarger, H.E. (1980) Nucleotide sequence of *ilvGEDA* operon attenuator region of *Escherichia coli*. *Proc. Natl. Acad. Sci. 77*, 1823–7.

Nixon, S.E., and Mawer, G.E. (1970) The digestion and absorption of protein in man. 2. The form in which digested protein is absorbed. *Br. J. Nutr. 24*, 241–58.

Novick, A. (1958) Some chemical bases for evolution of microorganisms. In A.A. Buzzati-Traverso (ed.) *Perspectives in Marine Biology*, pp. 533–546. Berkeley: University of California Press.

Novick, A., and Weiner, M. (1959) The kinetics of beta-galactosidase induction. In R.E. Zirkle (ed.) *A Symposium on Molecular Biology*, pp. 78–90. Chicago: The University of Chicago Press.

Portalier, R. Robert-Baudouy, J., and Stoeber, F. (1980) Regulation of *Escherichia coli* K-12 hexuronate system genes: *exu* regulon. *J. Bacteriol. 143*, 1095–107.

Pouyssegur, J., and Stoeber, F. (1974) Genetic control of the 2-keto-3-deoxy-D-gluconate metabolism in *Escherichia coli* K-12: *kdg* regulon. *J. Bacteriol. 117*, 641–51.

Ptashne, M., Jeffrey, A., Johnson, A.D., Maurer, R., Meyer, B.J., Pabo, C.O., Roberts, T.M., and Sauer, R.T. (1980) How the λ repressor and *cro* work. *Cell 19*, 1–11.

Raibaud, O., and Schwartz, M. (1984) Positive control of transcription initiation in bacteria. *Annu. Rev. Genet. 18*, 173–206.

Rigby, P.W.J., Burleigh, B.D., and Hartley, B.S. (1974) Gene duplication in experimental enzyme evolution. *Nature 251*, 200–4.

Roberts, J.W. (1969) Termination factor for RNA synthesis. *Nature 224*, 1168–74.

Sadler, J.R., and Novick, A. (1965) The properties of repressor and the kinetics of its action. *J. Mol. Biol. 12*, 305–27.

Savageau, M.A. (1969) Biochemical systems analysis II. The steady-state solution for an n-pool system using a power-law approximation. *J. Theoret. Biol. 25*, 370–9.

—— (1971) Concepts relating the behavior of biochemical systems to their underlying molecular properties. *Arch. Biochem. Biophys. 145*, 612–21.

—— (1972) The behavior of intact biochemical control systems. *Curr. Top. Cell. Reg. 6*, 63–130.

—— (1974) Genetic regulatory mechanisms and the ecological niche of *Escherichia coli*. *Proc. natl. Acad. Sci. 71*, 2453–5.

—— (1976) *Biochemical Systems Analysis: A Study of Function and Design in Molecular Biology*. Reading, Mass.: Addison-Wesley.

—— (1977) Design of molecular control mechanisms and the demand for gene expression. *Proc. Natl. Acad. Sci. 74*, 5647–51.

—— (1979a) Autogenous and classical regulation of gene expression: a general theory and experimental evidence. In R.F. Goldberger (ed.) *Biological Regulation and Development*, vol. 1, pp. 57–108. New York: Plenum Press.

—— (1979b) Growth of complex systems can be related to the properties of their underlying determinants. *Proc. Natl. Acad. Sci. 76*, 5413–7.

—— (1980) A systems theory of gene regulation. In *Proc. 24th Annual North American Meeting of the Society for General Systems Research with the American Association for the Advancement of Science*, pp. 125–133. San Francisco, CA.

—— (1983a) Models of gene function. General methods of kinetic analysis and specific ecological correlates. In H.W. Blanch, T. Papoutsakis, and G.N. Stephanopoulos (eds.) *Foundations of Biochemical Engineering: Kinetics and Thermodynamics in Biological Systems*, pp. 3–25. Washington, D.C.: American Chemical Society.

—— (1983b) Regulation of differentiated cell-specific functions. *Proc. Natl. Acad. Sci. 80*, 1411–5.

—— (1983c) *Escherichia coli* habitats, cell types, and molecular mechanisms of gene control. *Am. Nat. 122*, 732–44.

—— (1985a) Mathematics of organizationally complex systems. *Biomed. Biochim Acta 44*, 839–44.

—— (1985b) A theory of alternative designs for biochemical control systems. *Biomed. Biochim Acta 44*, 875–80.

—— (1985c) Coupled circuits of gene regulation. In R. Calander and L. Gold (eds.) *Sequence Specificity in Transcription and Translation*, pp. 633–642. New York: Alan R. Liss.

Savageau, M.A., and Voit, E.O. (1987) Recasting nonlinear differential equations as S-systems: a canonical nonlinear form. *Math. Biosci. 87*, 83–115.

Schibler, U., and Sierra, F. (1987) Alternative promoters in developmental gene expression. *Annu. Rev. Genet. 21*, 237–57.

Smith, G.R., and Magasanik, B. (1971) Nature and self-regulated synthesis of the repressor of the *hut* operons in *Salmonella typhimurium*. *Proc. Natl. Acad. Sci. 68*, 1493–7.

Sorribas, A., and Savageau, M.A. (1989) A comparison of variant theories of intact biochemical systems, 1. Enzyme – enzyme interactions and biochemical systems theory. *Math. Biosci.* (in press).

Squires, C.L., Lee, F.D., and Yanofsky, C. (1975) Interaction of the *trp* repressor and RNA polymerase with the *trp* operon. *J. Mol. Biol. 92*, 93–111.

Voit, E.O., and Savageau, M.A. (1987) Accuracy of alternative representations for integrated biochemical systems. *Biochemistry 26*, 6869–80.

Webster, G., and Goodwin, B.C. (1982) The origin of species: a structuralist approach. *J. Social Biol. Struct. 5*, 15–47.

Whipp, M.J., and Pittard, A.J. (1977) Regulation of aromatic amino acid transport systems in *Escherichia coli* K-12. *J. Bacteriol. 132*, 453–61.

Wright, S. (1931) Evolution in mendelian populations. *Genetics 16*, 97–159.

Yanofsky, C. (1981) Attenuation in the control of expression of bacterial operons. *Nature 289*, 751–8.

6. Origins of order in evolution: self-organization and selection

STUART A. KAUFFMAN

This article is written as prolegomena, both to a research programme, and to a forthcoming book discussing the same issues in greater detail (Kauffman, in preparation). The suspicion that evolutionary theory needs broadening is widespread, but to accomplish this will not be easy. The new framework I discuss here grows out of the realization that complex systems of many kinds exhibit high spontaneous order. This implies that such order is available to evolution and selective forces for further moulding. It also implies, quite profoundly, that the spontaneous order in such systems may enable, guide and *limit* selection. The spontaneous order in complex systems therefore, implies that selection may not be the sole source of order in organisms, and that we must invent a new theory of evolution which encompasses the marriage of selection and self-organization.

The overall themes I wish to explore are these:

1. Complex systems, such as the genomic regulatory networks underlying ontogeny, exhibit powerful 'self-organized' structural and dynamical properties.

2. The kinds of order which arises spontaneously in such systems is strikingly similar to the order found in organisms.

3. This raises the plausible possibility that the spontaneous order found in such complex systems accounts for some or much of the order found in organisms.

4. The existence of strongly self-ordered properties in complex systems implies that selection must be acting on systems with their own 'inherent' properties; hence, at a minimum, what is ultimately selected may often reflect a compromise between selection and the spontaneous properties of the class of systems upon which selection is acting.

5. Such compromises reflect the fact that selection is in general a combinatorial optimization process in a rugged fitness landscape with many peaks, ridges and valleys. The typical structure of such landscapes, and the population flow upon them under the drives of mutation and selection, ensure that attaining and maintaining high optima, with properties which are very rare in the class of systems under selection, cannot usually occur.

6. The typical failure of selection to be able to *avoid* the typical properties of the class of systems under selection is therefore expected to sustain points 3 and 4 above. The spontaneous properties of complex systems under selection will often be similar to those generic in the class of systems under selection, not *because* of selection, but *despite* it.

7. In turn, if many features of organisms reflect the generic properties of an entire class of systems under selection, not the particular successes of selection, then non-reductionistic theories based on analysis of those generic properties can be expected to be predictive of features of organisms. We should, in short, be able to predict many features found in organisms without needing to know the details.

8. The capacity of selection to achieve rare highly functioning forms is governed by the statistical features of the adaptive landscape over which selection tries to pull an adapting population. As we shall see, many landscapes are 'bad' in the sense that adaptation is likely to become trapped on mediocre local optima which become ever more mediocre as the complexity of the entitities under selection increases. We must envisage a yet further broadening of evolutionary theory to encompass the possibility that selection has achieved entities with the internal properties which allow them to adapt on 'good' adaptive landscapes with high optima. The initial conditions for this to occur will also be discussed.

In preliminary summary: complex systems are self ordered, promising relief to selection as the sole source of order in biology. The relief must be purchased, however, by a broadened theory which considers the capacity of such systems to adapt on rugged fitness landscapes, and the capacity of selection to alter the kinds of enties which exist, hence the kinds of landscapes they evolve upon. Evidently, we must take Darwin's central profound idea profoundly seriously, but move beyond it.

This article is organized as follows. In the following section I introduce the concept of an adaptive landscape, and characterize briefly the properties of adaptive walks to local optima in rugged landscapes. In the next section I introduce a class of models which generate a family of 'tunably' rugged landscapes and seek generic statistical features of adaptive walks on such landscapes. In the next section I discuss briefly the emergence of spontaneous order in models of genomic regulatory systems and the parallels between the generic properties of this class of systems and features of ontogeny. In the next section, I report briefly on the capacity of selection to act on genomic regulatory systems, and the limitations due to the statistical features of their rugged landscapes. In the final section I return to the issue of whether selection may achieve entities which have the internal properties allowing them to adapt on 'good' fitness landscapes, and discuss the extent to which this may circumvent or modify the general theme – that the selection cannot avoid the typical properties of the class of systems upon which it operates.

RUGGED ADAPTIVE LANDSCAPES

The adaptive landscape metaphor in evolutionary biology is at least as old as Wright (1932). The metaphor is limited in certain respects. Thus, it is well known in population genetics that a population undergoing selection on two or more loci, with two or more alleles per locus, may not flow 'up' the fitness landscape due to recombinational constraints between the two loci. Density and frequency dependent effects may also limit the simplest vision of a fitness landscape, where fitness is a function of genotype and attendant phenotype (Ewens, 1979). More generally, the fitness of an organism is a function of its own phenotype and the number of others of the same or other phenotypes, hence not a pure function of one phenotype alone.

Despite these and other limitation, the fitness landscape image is powerful, basic, and a proper starting point to think about selection. We conceive, next, of a very simple space of objects, namely peptides, and analyse the character of adaptive walks via one-mutant fitter variants to local or global optima for a defined functional property of the peptides (Eigen, 1985; Smith, 1970).

Sequence space

Consider the set of all peptides of length 10. With twenty amino acid types, there are twenty-to-the-ten possible peptides with ten amino acids. Define the one-mutant neighbours of a peptide to be all those sequences which can be obtained by changing one amino acid to one of the nineteen other possibilities. Thus, each peptide has $19N = 190$ one-mutant neighbours. We are then conceiving of a peptide sequence space with all twenty-to-the-ten possible peptides of length 10, ordered in a high-dimensional sequence space in which each peptide is a point, and is connected by a line to its 190 one-mutant neighbours. This space, in short, is a sequence space. Distance between points along connected lines passing through other points correctly represents the minimum number of changes in amino acids to convert one sequence to another.

We then need to define a fitness landscape. We do so based on the capacity of each peptide to perform some function. For example, we might measure the affinity with which each peptide binds a specific hormonal receptor on a cell surface. This measured affinity then can be thought of as a measure of the 'fitness' of each peptide with respect to that function. Because each peptide has a measured affinity, and the peptides are arranged as points in an ordered sequence space, the measured affinities constitute a fitness landscape over the sequence space.

Consider next the simplest possible caricature of an adaptive walk in peptide space. We imagine that the adaptive process begins with some arbitrary peptide. Next we imagine that at each generation one or more mutant variants of that peptide are produced. In the simplest case, a single-mutant variant is produced, and it is mutant in a single amino acid. This one-mutant neighbour may be fitter than the initial peptide. If so, let the adaptive process step to this improved variant. If not, let the process remain 'at' the initial peptide and try another one-mutant variant. (I show below that this simplest case corresponds to a well-defined reasonable population-genetic situation.)

In this simplest image, an adaptive *walk* begins at an arbitrary peptide, and, on each trial, samples a one-mutant variant of the current peptide, and 'moves' to that variant only if fitter. Thus, the process is constrained to pass via one-mutant variants until it reaches a peptide which is fitter than all its one-mutant neighbours.

Obviously, the character of an adaptive walk depends upon the disposition of the fitness values in the fitness landscape. That distribution might range from very smooth, such that neighbouring peptides have highly similar fitness values, to rugged landscapes with many peaks, ridges and valleys, but still have substantial correlation between the fitness values of one-mutant neighbours, to fully uncorrelated landscapes in which the fitness values of one-mutant neighbours were completely unrelated to each other.

Given any fitness landscape, the immediate natural questions which arise are:

1. How many improvement steps are taken on an adaptive walk before coming to rest on a local optimum?

2. How many local optima exist?

3. How many alternative local optima are accessible via branching adaptive walks from an initial arbitrary peptide?

4. How does that alter with the fitness of the initial peptide?

5. How 'fit' are the local optima with respect to the mean fitness in the space of peptides?

6. How do these properties depend upon the ruggedness of the landscape and the complexity of the entities (here peptides) under selection

Given any fitness landscape, the actual flow of a population across it depends not only upon its structure, but the population size, mutation rate, initial dispersion across the landscape, and other factors (Ewens, 1979). Thus, even if we understand the structure of such landscapes, we still will not understand population flow upon them.

More complex questions then arise. What kinds of entities, peptides or otherwise, have given types of fitness landscapes and why? Are some landscapes easier to evolve upon than others? Can selection achieve entities which 'live on' good landscapes? I discuss these below.

The fully uncorrelated landscape

Simon Levin and I have recently analysed the case of adaptive walks on fully uncorrelated landscapes (Kauffman and Levin, 1987). These are constructed by assigning each peptide a fitness value drawn at random from a fixed underlying fitness distribution. To be concrete, we may take that distribution to be uniform between 0.0 and 1.0. Since the adaptive process we are considering passes via one-mutant fitter variants, the actual fitness values are not important, and can be replaced by the *rank order* of the peptides, from worst to best.

We have established the following features of adaptive walks on uncorrelated landscapes:

1. The number of local optima is very large. For peptides with twenty amino acids it is 20-to-the-N/(19N + 1). That is, the expected number of local optima is just the number of entitities in the space divided by one plus the number of one-mutant neighbours to each entity.

2. Walks to local optima via fitter one-mutant variants are very short: log 2 (19N). That is, the number of fitter variants encountered on a walk is just the logarithm base two of the number of one-mutant neighbours to any peptide in the space. This implies that, for peptides of length 10, walk lengths are on the order of 7 or 8, while for proteins of length 100, walks are of the order of length 11.

3. Only a small fraction of all local optima are accessible from a single initial peptide via branching adaptive walks.

4. As fitness increases at each step along an adaptive walk, the number of fitter one-mutant neighbours decreases by a half. Thus, branching walks to alternative optima initially have many alternative routes upward, and these dwindle to single routes which wend upward to local optima. Such branching trees, in short, are bushy at the base and dwindle to single lineages.

Brief justification

I have idealized a walk as sampling one-mutant neighbours one at a time, and constrained to pass via fitter neighbours. This idealization corresponds quite closely to a population with a low mutation rate, in which the rate of finding a fitter variant is very low, while the fitness differentials between improved variants and 'wild type' is moderate. In this case, on a slow time-scale, the population uncovers a fitter mutant, and on a fast time-scale the fitter mutant sweeps through the population. Thus, roughly, the population 'hops' to a fitter one-mutant neighbour. Gillespie (1974) has shown that this limiting case corresponds to a continuous time, discrete-state Markov process, with the population passing from one state to a neighbouring fitter state.

Obviously, in reality, a population may harbor one-mutant, two-mutant, . . . J-mutant variants at the same time. Actual population flow upon a landscape is more complex that this simplest case, which I have adopted as a means to analyse the structure of the landscape itself, rather than flow upon it.

Universal features of long-jump adaptation

Real populations may be able to 'jump' long distances in peptide spaces in single mutational events. For example, recombination is a mutational process which may substitute a large number of amino acids simultaneously into a peptide or protein. Such a mutation can be thought of as jumping long distances across peptide space.

Consider next any correlated but highly rugged landscape, such as the Alps. If a mutational process jumps *beyond* the correlation length of that landscape, then the fitness of the point reached is fully uncorrelated with the fitness of the point left. This corresponds to the fact that altitudes of points fifty kilometres from any point in the Alps are essentially uncorrelated with the altitude of the point left. Thus, long-jump processes in correlated but rugged landscapes encounter an uncorrelated landscape.

Levin and I (Kauffman and Levin, 1987) established several universal features of such adaptive processes:

1. After each improved variant is found, the waiting time to find the next improved variant *doubles*. Thus, the rate of improvement slows rapidly. The consequence is that the expected number of improvement steps, S (after G long-jump trials), is $S = \log_2 G$.

2. As is the case described above, the number of alternative fitter long-jump variants is halved on average at each improvement step, thus branching lineages are bushy at the base and the rate of branching dwindles as fitness increases.

A complexity catastrophe limits selection as complexity increases

We have uncovered a fundamental and previously unexpected limitation of adaptation on rugged landscapes via long jumps. The same limitation applies to adaptation via one or few mutant variants on fully uncorrelated landscapes. As the *complexity* of the entities under selection increases, there is a marked tendency for the *attainable local optima* or the states attained after long jump walks of any fixed length to *fall* toward the mean fitness of the space of entities! In the current case, the complexity of the entities is just the length of the peptides. As N increases, the number of local optima increases, and the number of one-mutant or J-mutant variants of each peptide increases, but the actual fitness of local optima, or fitness attained after a fixed walk length, *decreases*. This constitutes a kind of *complexity catastrophe*.

This limitation is terribly important. We have already assumed in our idealization that selection is *always strong* enough to pull a population to any optimum and hold it there, an assumption which is generally false. Even with this assumption, this new limitation says that as entities under selection become more complex, then, in the limits of uncorrelated landscapes and walks via neighbouring fitter variants, or for rugged landscapes and long-jump adaptational walks, the fitness achieved *falls*. More complex means more mediocre. This is a powerful limitation. We must ask whether there may be conditions which circumvent this limitation. Those conditions would be upon the character of the *adaptive landscape itself*. We must ask whether there are conditions such that entities under selection can become more complex, yet the adaptive peaks do not dwindle. The answer is 'yes', as we discuss in the next section.

GENERAL EPISTATIC INTERACTIONS AND RUGGED LANDSCAPES

What properties of complex systems control the ruggedness of the fitness landscapes upon which they may evolve? Do peptides adapt on highly correlated, or very rugged landscapes? What of large proteins? What of the coupled genomic regulatory system where the activities of structural genes are regulated by a variety of control genes in complex circuitry? What of the morphology of an organism? In general, the answers are unknown. In fact, to my knowledge, the question of the relation between the character of entities under selection and the ruggedness of their adaptive landscapes has never been addressed at all.

To approach this issue, I here introduce a general model of 'epistatic' interactions among traits in an organism. This model is similar in spirit to spin glasses (Anderson, 1985) and, as we shall see, appears to have wide applicability.

Consider an organism with N 'traits'. For the moment let us restrict attention to the case where each trait is simply present or absent, denoted by a 1 or 0. With N traits, there are therefore two-to-the-N possible combinations of traits which might be present or absent. I wish to consider the fitness contribution of each of the N traits. In general, the fitness contribution of each trait may well depend upon the presence or absence of others of the N traits. Such dependencies are called epistatic interactions in population genetic models where one thinks of the fitness contribution of the gene at one locus, which may depend not only on the allele of that gene, but depend epistatically on the alleles present at some other loci. Because I wish to study the effect of *richness* of epistatic interactions on the ruggedness of fitness landscapes, I shall assume that the fitness contribution of each trait depends upon itself and K other traits among the N. Increasing or decreasing K then alters how many traits affect the fitness contribution of each trait. In addition to the constraint that each trait's fitness contribution depends upon the presence or absence of K other traits, I shall add one further constraint for the moment. I assume that if trait 'i' matters to trait 'j', then trait 'j' matters to trait 'i'. Within these constraints, I shall assign the K traits mattering to each trait, i, randomly.

The fitness contribution of each trait, i, thus depends upon the presence or absence of itself plus K other traits. Therefore, the fitness contribution of the i-th trait must be specified for each of the two-to-the-(K + 1) combinations of presence or absence of the K + 1 traits which affect trait i. To do so, I shall draw a different random number from a uniform distribution between 0.0 and 1.0 and assign it to each of the two-to-the-(K + 1) combinations of traits. The same is done for all N different traits. Therefore, for any particular combination of N traits being present or absent, the fitness contribution of all N traits is specified. The fitness of the entire combination of N traits is defined as the sum of contributions of all traits, divided by N. This yields an overall fitness for each combination of traits which lies between 0.0 and 1.0.

The model I have defined is a random epistatic fitness model, tuned by N and K. It is open to many interpretations. For example, the 'traits' might instead be considered amino acids in a peptide. In the present case we would conceive of only two types of amino acids. Given a spatial interpretation of the N bits as amino acids, we might choose the K which matter to each site from neighbouring or from random sites in the peptide. Obviously the model can be generalized to twenty rather than two kinds of amino acids. Under the interpretation that the 'traits' are genes, each open to two or many alleles, the model is interpretable as a model of epistatic interactions of K loci upon each site, with the K distributed in some defined order on neighbouring or random sites on the chromosome. This genetic interpretation corresponds to a haploid model with a single copy of each chromosome.

Given N, K, and the distribution of K among the N, what do the resulting fitness landscapes look like? Return to the simple case where each 'trait' is present or absent. Then the space under consideration is the ordered N dimensional Boolean hypercube, where each vertex or point is a particular combination of the presence or absence of each of the N traits. Further, each point is a one-mutant neighbour of N other points, attained by changing one trait from present to absent, or vice versa. Each point has a well-defined fitness. Therefore, a well defined fitness landscape exists over the Boolean hypercube.

K = 0 corresponds to a fully correlated landscape with one optimum

Consider the case where K = 0. Here the fitness contribution of each 'trait' is independent of all other traits. Since the fitness of the entire system is just the sum of the fitnesses of the N traits, this model is identical to a simple additive fitness population genetic model with N genes, each with two alleles and no epistatic interactions among the genes. In the present case, each trait, by chance, makes a higher fitness contribution if present, or if absent. Thus a single global optimum exists in which each trait is in its more valuable state, present or absent. Note that any other combination of traits is less fit than the global optimum. Further, any such suboptimal combination of traits lies on a connected pathway via fitter variants leading to the single (global) optimum. Each trait in its less valuable state need merely by 'flipped' to its more valuable state. Note next that the fitness of one-mutant neighbours will be nearly the same. This follows since flipping a single 'trait' will, on average, change the total fitness less than $1/N$. That is, the $K = 0$ landscape is a highly *correlated* landscape.

Deeper insight into the present $K = 0$ case is readily attained. Because each fitness contribution has, in the present case, been drawn at random from a uniform distribution between 0.0 and 1.0, and the fitter value is the fitter of two random draws, on average the less fit value is 0.333 and the more fit value is 0.666. In a system with N 'traits', the expected fitness of the global optimum is 0.6666. Furthermore, the expected fitness of the global optimum is obviously *independent* of N. As N increases, the fitness of the global optimum remains .66666. If an adaptive walk begins with an arbitrary combination of traits, its fitness will be about 0.5. Half the traits will be in their less valuable state, thus on the order of $N/2$ improvement steps occur on an adaptive walk to the single and global optimum. Along that walk, the number of fitter neighbours decreases by one on each step upward.

$K = N$ corresponds to a fully uncorrelated fitness landscape

At the opposite extreme, suppose the fitness contribution of each trait depends upon all traits. Then $K = N - 1$, or, for large N, K effectively equals N. Consider a particular combination of N traits. Change one trait from present to absent. Then the context of all N traits has changed. For each we shall have specified a fitness value at random from the uniform distribution. Therefore flipping a single trait yields a one-mutant neighbour whose fitness is an entirely new sum of N random samples from the uniform distribution. That is, the fitness values of one-mutant neighbours in the space are entirely random with respect to one another. The $K = N$ case therefore corresponds to a fully *uncorrelated* fitness landscape.

We already understand the character of adaptive walks on such a landscape. There are a very large number of local optima. Walks to local optima take on the order of $\log_2 N$, rather than $N/2$ for the $K = 0$ limit. The number of fitter neighbours is halved after each improvement step. From any initial combination of traits only a small fraction of the local optima are accessible.

Among the most critical features of $K = N$ landscapes is that they exhibit the *complexity catastrophe* described above. As the number of traits, N, increases, the fitness of attainable local optima *decreases* toward 0.5, the mean of the space.

0 < K < N – tunably rugged fitness landscapes

Clearly, increasing K from 0 to N, corresponding to increasing the epistatic interactions among the traits in the system, increases the ruggedness of landscapes from fully correlated to fully uncorrelated. The intent of the model now becomes clear, for the parameters N, K, and the distribution of K among the N, generate a *family* of expected landscapes with statistically characterizable adaptive walks. It may well be the case that this family covers a vast spectrum of 'real' adaptive landscapes which arise in a variety of combinatorial optimization problems from the 'travelling saleman' problem (Lin and Kernighhan, 1973), to walks in protein spaces during maturation of the immune response (Kauffman and Levin, 1987). At a minimum, this family of landscapes should offer initial intuition about the requirements on system construction and the ruggedness of the resulting landscapes.

To approach this issue, my colleague E. Weisberger and I have written a general computer program. I show the results of numerical Monte Carlo trials for different values of N and K, and for two different distribution rules on the K. In the first set we conceived circular peptides (to avoid end effects) and required the K sites bearing on each site to be its $K/2$ neighbours on either side. In the second set we removed the restriction that K be reflexive, and allowed each site to be affected by K randomly chosen other sites. The results do not appear to depend upon this difference strongly. Thus, N and K determine the major features of the class of landscapes.

The numerical results confirm the general expectations derived above:

1. Both sets confirm that for K = 0, the single global optimum has a fitness of 0.66666, and that the optimum is independent of N. Both confirm the expectation for the K = N case: that, as N increases, the fitness of local optima *decreases* towards 0.5. The complexity catastrophe on uncorrelated landscapes is real.

2. For K = 0, walk lengths are of the order of $N/2$, and the number of fitter neighbours declines by one per step. For K = N, walk lengths are of the order of $\log_2 N$, and the number of fitter neighbours declines roughly exponentially. For values of K between 0 and N, however, new features emerge.

1. For K small and fixed, for example, K = 2, 4 or 8; as N increases, the fitness of optima actually increases towards an apparent asymptote which becomes independent of N, and which is *larger* than the global optimum for K = 0. In other words, a small amount of epistatic interaction actually *helps create a landscape with optima of higher fitness!* And the fitness appears to become independent of N for large N. Thus, for K fixed (and perhaps small) complexity, the sense of increasing N does not lead to a complexity catastrophe. The space retains optima of high fitness.

2. What happens if K increases proportionally to N? Both sets show that for K = N/2, K = N/3, and K = N/4, as N increases the fitness of

optima first increase above the $K = 0$ case, then decline below it, and presumably fall toward 0.5 as N grows sufficiently large. Thus, if K is proportional to N, it appears that a complexity catastrophe occurs. Increasing complexity leads to landscapes with lower optima. K proportional to N would mean, intuitively, that as the number of traits in an organism (or number of amino acids in a peptide) increase, that a constant fraction of those traits affect the fitness contribution (i.e. function) of each trait or amino acid.

These last two observations carry the following implication. Consider K as a function of N. For K constant as N increases, the complexity catastrophe is averted. For K proportional to N, it is not averted. Thus for K equal to some monotonically increasing function of N, a transition must occur between landscapes which do not and do exhibit the complexity catastrophe. What that boundary is remains unknown.

How generalizable are these results? In the present simplest model we have sampled fitness values from a uniform distribution between 0.0 and 1.0. We do not yet know, but strongly conjecture, that the qualitative results are independent of the underlying distribution, which might be uniform, Gaussian, exponential, etc., so long as the fitness contribution for each combination of 'traits' is sampled from the *same* distribution. Indeed the results may generalize to a process which samples from a set of different distributions.

We will now summarize tentative conclusions. If, as N grows large, the number of 'traits' which impinge epistatically on any trait is bounded below some constant or slowly increasing value, then such systems adapt on rugged landscapes which avert the complexity catastrophe. Beyond some critical rate of increase in K, the complexity catastrophe sets in. I suspect that the implication of this general argument holds quite generally. For complex entities to evolve, the number of parts which directly impinge upon any part probably must grow very much more slowly than the number of parts in the whole. I turn in the next section to describe the self-organized dynamical behaviour of model genomic regulatory systems. There, it turns out that genetic networks in which each gene is directly affected by only a few other genes exhibit a marked order reminiscent of real ontogeny. In the subsequent section, I present evidence that genomic systems of low connectivity also adapt on correlated landscapes which avert the complexity catastrophe. Thus, the same property, low K, appears to abet orderly behaviour, and the capacity to evolve well!

SELF-ORGANIZED BEHAVIOUR IN GENETIC NETS OF LOW CONNECTIVITY

Metazoans such as mammals have on the order of 100,000 different structural and regulatory genes. Since Jacob and Monod discovered that products of one gene could activate or repress the activities of other genes, biologists have come to think of the genomic system as a kind of biochemical computer. The structural and regulatory genes are linked into some kind of circuitry, regulating and co-ordinating one another's behaviour. It is a canon of current developmental biology that all the diverse cell types of an organism contain the same set of genes co-ordinated by the same regulatory circuitry. While there are minor exceptions to this canon, it is close enough to truth for current purposes. In terms of this canon, the fundamental problem of cell differentiation is: why are cells different from one another, despite harbouring identical genomic systems? The answer, of course, is that cells differ because different subsets of genes are expressed in different cell types. It follows that it is the genomic circuitry and regulatory system which co-ordinates and engenders these different patterns of gene expression.

The presumptive complexity of a genomic system with about 100,000 genes regulating one another's activities bodes serious epistemological problems. Molecular genetics is fast uncovering the molecular machinery by which one gene acts on a second gene. Structural genes, which code for proteins, typically are flanked by specific DNA sequences which are bound by the protein products of other genes. This binding can turn on or turn off transcription of the structural gene to messenger RNA, thence to protein. Such adjacent DNA sequences, because they regulate adjacent genes on the *same* chromosome, are called *cis*-acting regulatory genes. The upstream genes whose products bind to cis-acting sequences are themselves called *trans*-acting, because they can exert regulatory influences on genes located on different chromosomes. Regulation is more complex than mere control of transcription. Different points in maturation of messenger RNA and translation to protein are also subject to regulation. For our purposes I shall lump all these into the term 'genomic regulatory network'.

The genomic regulatory network is subject to evolution in two quite different ways. First, of course, point and recombinational mutations may alter DNA sequences leading to new useful structural genes and proteins. Some of these altered proteins may themselves be trans-acting proteins, hence their alteration may alter the effect of the regulating gene upon its target genes. But beyond such mutations, chromosomal mutations literally shuffle genes from one to another position in the set of chromosomes. Such mutations include duplications, and various processes which disperse duplicated genes to new locations such as translocations, transpositions, inversions, conversions. Consider the obvious consequences. If gene A is adjacent to a cis acting sequence X, and a transposition moves a new gene B into position between X and A, then X may now control B as well as A. If B happens to be a trans-acting gene regulating a downstream cascade, then X now regulates the same cascade via B. In short, chromosomal mutants literally scramble the circuitry in the genomic regulatory system.

From the foregoing, it follows that *evolution of new cell types* occurs by modification of structural genes and of the 'wiring diagram' and 'logic' of the regulatory circuitry. I have attempted (1969; 1974; 1984) to study the genetic properties of the ensemble explored by evolution by exploring model genomic systems constructed at random within the constraints that there be N genes, each regulated by K other genes. Specifically, this means deciding at random which among the N genes are the K inputs to each gene, then deciding at random for each gene which among the two to the K possible Boolean functions of K inputs shall govern its behaviour. That is, both the wiring diagram and the logic are chosen at random. Once chosen, the structure is fixed.

The results of many numerical simulations for different values of N and K have shown that for small K, for example K = 2, such networks exhibit powerful order which is strongly reminiscent of real cells. Briefly, such a network is a finite automaton. Each combination of activities of the N genes is a state. Thus there are two to the N states. At each moment, the net is in a state. The genes each examine the activities of their inputs, consult their switching rule, then all genes synchronously assume their designated new activity value. Thus the net passes from one state to another state. Over time, the system passes through a succession of states. Since there is a finite number of states, eventually the system re-enters a state previously encountered. Thereafter, since the net is deterministic, it cycles repeatedly through this re-entrant cycle of states. It is critical that many states may lie on sequences of states which flow to the same state cycle. But the network may harbour more than one state cycle. If so, some states flow to one of these asymptotic attractors, while others flow to the other state cycles. Wherever the system is released, it will flow to one state cycle attractor. Thus the system comes to cycle about one or another of its state cycles.

The order which emerges spontaneously when K = 2 includes the fact that the lengths of state cycles are small, of the order of square root N, the number of states cycles in a net's behavioural repertoire is also about root N, each cycle is stable to transient reversal of the activity of most genes one at a time, and, if unstable, will flow to only a few of the remaining state cycles.

I have for a number of years wanted to see in this spontaneous order a deep similarity to real ontogeny. I make a single interpretation: I identify a *cell type* with a *state cycle attractor* of a genomic network. That is, a cell type is a stable recurrent pattern of gene expression engendered by the genomic logic. Given this interpretation, the theory makes a number of predictions which are surprisingly accurate. Thus, the number of cell types in an organism should increase as a square root function of the number of genes! This is surprisingly close to what is observed across many phyla (Kauffman, 1969; Kauffman, 1974; Kauffman, 1984). If a cell type is a state-cycle attractor, then differentiation is the passage from one attractor to another. These models imply that any cell type can only flow to a *few neighbouring cell types*, and from thence to a few other cell types. But this implies that ontogeny should occur along *branching developmental pathways*. Indeed, all contemporary multicellular organisms develop along just such branching cell lineage pathways. Presumably this has remained true for the past 600 million years. I return to this below, for I shall want to ask whether this feature of organisms reflects selection, or is such a deep property of genomic systems that it is *impervious* to selection.

This class of models predicts many other features of current organisms, such as the similarity in gene expression patterns in different cell types (about 90 per cent to 95 per cent overlap), the existence of a large core of genes active in all cell types, the typical cascading consequences of deleting a single gene, and so forth (Kauffman, 1969; 1974; 1984).

Without belabouring the detailed results, the main point to cull from all these studies is that even randomly constructed model genomic systems, with few inputs per gene, exhibit an order which is strongly reminiscent of that seen in ontogeny. This implies that our intuition about complex systems is wrong. Order emerges spontaneously. From this it follows that this order may account for the origin and persistence of such order in organisms. The spontaneous order is a least a handmaiden to selection, but of course, selection operates continuously. This brings us to the central problem, which I believe faces evolutionary theory. We need to broaden evolutionary theory to understand how selection can act upon, with and through systems which have their own strongly self-organized properties. How is selection enabled, guided and limited by those properties? Note that no area of science has dealt with this problem. In the next section, I discuss the initial steps to be taken in this direction.

SELECTION AND ITS LIMITS FOR DESIRED CELL TYPES

Can selection, operating both on structural genes and the 'circuitry and logic' of a genomic regulatory system, mould such a regulatory system to achieve arbitrary 'good' cell types? Obviously, to broach this question first requires some model of a cell type, and the ways in which mutation and selection can modify cell types. The ensemble of genomic network models described above provides just such a framework.

To be concrete, I shall ignore the evolution of new structural genes coding for new useful proteins with novel enzymatic, structural or other features. Rather, I shall focus on evolution of the circuitry and logic to alter the co-ordinated patterns of expression among a constant set of structural genes. Within that limitation, and in the framework of the ensemble theory developed above, evolution can proceed via mutations in the regulatory *connections* between genes, thus altering the 'wiring diagram' of the genomic circuitry; or evolution can proceed by altering the regulated behaviour of a gene as its inputs alter activities. That is, mutations can affect the Boolean function characterizing the response of any regulated gene. To be concrete, chromosomal mutations may alter the wiring diagram, or such mutations and point mutations may alter the local rule. For example, a mutation which prevents a repressor protein from binding to its *cis* acting site may render an adjacent regulated gene constitutively (constantly) active. The gene now realizes the 'tautology' logical function, always active.

Evolution explores an ensemble of regulatory networks

In the second and third sections of this article I discussed the structure of fitness landscapes, where the points in the space were conceived to be proteins, each a one-mutant neighbour of those other proteins obtainable by altering one amino acid in the protein's primary sequence. The concept of rugged fitness landscapes is far broader, however. Consider a genetic network with N genes, and K inputs per gene. Each such network is a member of a vast ensemble of all networks constructable with N and K as constraints. Each network is a one-mutant neighbour of all networks which can be attained by altering one regulatory connection, or one 'bit' in the Boolean function regulating one gene. Therefore, we can, again, define the concept of a high-dimensional space, where each point is an entire genetic regulatory network, and its one-mutant neighbours are all those genetic networks accessible by minimal changes of wiring diagram or logic. Thus, we can consider evolution as occuring across a space of genetic regulatory networks.

Any property of such networks might serve as a property upon which selection acts. To be concrete, let us define the fitness of a genetic regulatory system by how closely the patterns of gene activity occuring on one or another of its state cycles match to an arbitrary pattern of gene expression. That is, we shall define the 'fitness' of a genetic network of N genes by how closely one of its cyclic attractors comes to having a pattern matching an arbitrary 'desired' or 'good' pattern of gene expression across the N genes. For example, if the arbitrary pattern among the N genes is (101010101010 . . .), then the fitness of a given network is given by the fraction of genes whose activities match this pattern on the best matching state of one of the state-cycle attractors. This therefore assigns a fitness between 0.0 and 1.0 to each network, and hence generates a fitness landscape over the space of genomic systems.

As before, this fitness landscape may be more or less rugged. Also as before, in the simple case of adaptive walks constrained to pass via fitter one-mutant variants, we can ask how many steps occur on the way to local optima, the number of alternative local optima accessible, the similarities of those optima. Of course we are interested in how closely such optima match to the arbitrary 'good' pattern for which we are selecting. That is, can selection achieve *arbitrary good* cell types? And we are interested, as the complexity of the genomic system, N, increases, to know whether selection leads to less fit optima. Does the complexity catastrophe creep in?

I note in passing that these selection studies are a root form of learning in massively parallel processing systems. Thus, in the neurobiological context, models of distributed-content addressable memories use attractors as internal memories of external events (Kauffman and Smith, 1986). Asking whether mutation and selection can achieve desired attractors is a form of asking whether mutation and selection can be used to evolve parallel processing networks to have desired 'memories' (Hopfield, 1982). It is a further point of interest that any such system, if placed in a varying environment, will spontaneously classify its environments into equivalence classes given by those environments which leave a given attractor unaltered. In other words, these kinds of model genomic systems, if placed in an environment, naturally classify the *same* environment in different ways according to which attractor the system attains, and classify different environments as the *same* if those environments leave a given attractor unaltered. Non-linear systems with multiple attractors exhibit the basis of classification and a kind of cognition of their environments. Selective evolution for useful behaviours is thus a start towards evolving systems which classify their environments, recognize and act upon them. For other ideas in this area, see the article by Varela *et al.* in this volume.

Numerical studies of K = 2 and K = 10 Boolean networks

The general results of numerical simulations of such adaptive walks are these:

1. For Boolean networks with N = 50 and N = 100, and K = 2 or K = 10, the first major conclusion is that selection never achieves networks with fitness 1.0. Rather, fitness increases from an initial 0.5 and typically stops at a local optimum with values of about 0.65 or 0.7. This means that, in general, adaptation by one-mutant neighbours becomes trapped on local optima far below the logically possible perfect network. This is not abetted significantly by allowing two-mutant or five-mutant variants to be sampled.

2. The second general result is that the rate of finding fitter variants slows rapidly. This implies that, as fitness increases, the fraction of fitter neighbours dwindles rapidly.

3. The third general result is that *long-jump* adaptation, in which up to half the connections or one fourth the bits in Boolean functions are altered at once, reveals the expected general features of long-jump adaptation on rugged landscapes. Thus, the waiting time to find fitter variants doubles on average after each improvement step. And as N increases, the fitness attained after any fixed number of tries *falls* towards the mean fitness in the space of systems, 0.5.

4. The fitness landscape for $K = 2$ is highly correlated. One-mutant alterations to connections or logic make little difference to fitness. By contrast, the fitness landscape for $K = 10$ networks is far more rugged. Altering one connection, and even one bit in a Boolean function, typically alters a net's fitness drastically. Thus, $K = 2$ nets adapt on a more correlated landscape than do $K = 10$ networks.

5. Given the conditions in point 4 above, it is interesting to ask if $K = 2$ networks exhibit the complexity catastrophe for adaption via one-, two-, or five-mutant neighbours. (They do for long jump adaptation, of course). The answer is *no*. As N increases, the fitness of local optima remains essentially constant. The results for $K = 10$ networks are not available.

Point 5 is particularly interesting. Recall from the third section that for that NK spin-glass-like model, we found that for K small, optima were insensitive to N, while for $K = 0.5 N$ or $K = N$, the complexity catastrophe sets in as N increases. Thus there and here, low connectivity yields a correlated landscape, and the capacity to avoid the complexity catastrophe as complexity N increases. I stress, of course, that the two models are not at all identical. In the third section, the NK models are taken as general epistatic models. In the fourth section, the NK models are *dynamical models* of Boolean switching networks with N genes, and K inputs per gene. Landscapes in the switching networks are defined by attractors. Nevertheless there may be deep homologies between them with respect to ruggedness of landscape.

A major conclusion of both NK models, then, is that the complexity catastrophe can be averted by systems having the proper properties to adapt on correlated landscapes where optima do not wither toward the mean as complexity increases. In both cases this seems to require low connectivity among a rich system of many interacting parts. I turn in the next section to wonder if selection can select for systems which adapt on 'good' landscape.

Two even more fundamental observations about these simulations must be stressed. First, the fact that strong selection (i.e. selection which is always able to pull an adapting population 'to' a fitter variant) becomes trapped on local optima far from the logically possible Boolean network which matches a given pattern of gene expression precisely, very strongly implies that real selection acting on real organisms *cannot* sculpt arbitrary patterns of gene expression as the cell types of an organism. Whatever our intuitions have been, cell types almost certainly are not 'precise' in that sense. Evolution must make do with local optima, the best attainable on adaptive walks from whatever the starting point may be. The 'developmental program' beloved of developmental biologists is almost certainly not able to evolve to arbitrary logics from any given starting point.

Second, and perhaps most fundamentally, selection in rugged landscapes typically cannot *avoid* the typical features of members of that landscape. Adaptive walks in a space of genomic regulatory systems: (1) become stopped on local optima; (2) in the limit of long-jump adaptation climb higher at an ever slower rate remaining far from global optima; or (3) in the face of a high mutation rate which disperses a population away from attained optima, wander ergodically among sub-optimal genomic systems. In all cases it appears that selection is limited in its capacity to attain genomic systems, with features which are extremely untypical of the space of genomic systems in which adaptive evolution is occuring. If this is true, it bears the deepest consequences. For, if selection can typically not *avoid* the typical or generic features of the genetic regulatory systems in the ensemble explored, then those typical features should 'shine through', and be found in organisms, not because of selection, but despite it. Here then is a major theme. If selection cannot avoid the generic features of complex genomic regulatory systems which are members of some definable ensemble, then those generic properties may prove to be biological universals. Their explanation would not lie in details of structure, nor details of selective forces in branching phylogenies, but in mere ensemble membership. As phase transitions in water are understood as a form of critical phenomena in a general class of systems without precise accounting for locations of water molecules at the transition, so some or many ordered properties of organisms may find their explanation in the generic properties of the ensemble of which the organisms are members.

In view of this, I now return to the fact that model genomic systems with few inputs per gene have the property that each state-cycle attractor which models a cell type is a 'neighbour' of only a *few* other model cell types. Thus, this broad class of genomic models predicts that any cell type can differentiate into only a few other cell types, and they in turn to a few others. As noted earlier, this implies that ontogeny must be based about *branching* cell differentiation pathways, as is indeed the case. Does this organization of ontogeny, presumably present since the late Precambrian, reflect an achievement of selection? Or is it a property which is so deeply embedded in the entire ensemble of genomic regulatory systems accessible to selection, so deep a property of parallel processing nonlinear dynamical systems, that selection cannot avoid this property? I suspect the latter. And add that it is no trivial property, for it is the same property which assures the homeostatic stability of cell types in the face of perturbations, which ensures that a variety of specific and non-specific 'inductive' agents induce the *same* differentiation transformation, hence which underlies the subsequent elaboration of the logic of development. The possibility that such branching pathways reflect self-organization and not selection, is, I suspect, the harbinger of many possible new universals reflecting the balance between self-organization and selection.

All of this, however, brings us to an even tougher issue. Is selection doomed to wander among systems in a given ensemble? Or can selection actually change the way genomic systems are constructed, hence change the ensemble being explored, its generic properties, and the character of fitness landscapes upon the ensemble? Can selection find ensembles with 'good landscapes'?

SELECTION FOR 'GOOD' LANDSCAPES

What kinds of systems have 'good' landscapes? It is clear that fully uncorrelated landscapes are not 'good' in at least two senses:

1. First, in a correlated landscape, advantage may be taken of the correlation structure to 'look' where the looking is good. For example, many landscapes may prove to be self-similar. Spin-glass energy landscapes are an example (Anderson, 1985), and my first NK models in the third section may also be self-similar. That means that the landscape is fractal, with small hills and local optima located in the 'sides' of similarly-shaped, but wider and higher hills and optima, which in turn are on the sides of still larger, wider hills with higher optima. In such a correlated landscape, not only does it pay to look in the vicinity of modest local optima by jumping just far enough away to avoid trapping, but it may be the case that local optima carry global information about the general region of space with yet higher optima. Thus, one optimum is a good starting place to find yet higher optima. In contrast, in a fully uncorrelated landscape, locations of optima can carry no information about good regions of space. The search may as well be entirely random.

2. In fully uncorrelated landscapes, the complexity catastrophe creeps in. As systems become more complex, the optima attained relapse towards the mean fitness in the ensemble being explored.

In contrast, we have so far encountered only one class of entities which adapt on landscapes which avoid these problems. Such entities appear to have the property than any component directly interacts with a small number of other components in the system, and that small number remains bounded small in some as yet unknown way as the number of components in the system increases. Can selection 'tune' K and keep it small?

Let us consider two different levels of systems upon which selection may act. Consider proteins first. Small peptides (i.e. sequences of amino acids with about twenty or fewer members) probably adapt on very uncorrelated landscapes, for substitution of any single amino acid seems likely to have marked effects on the behavior of the peptide. By contrast, in large proteins, many amino acid substitutions have little effect on the overall architecture and function of the protein. Thus, large proteins almost certainly evolve on more correlated landscapes than small peptides. Evolution has opted for large proteins, by and large. This may reflect the commonly held view that large proteins can fold better, hence form better enzymes and structural components of cells. But it may also reflect the fact that in selecting for large proteins, selection has also achieved the kinds of entitities which adapt on more correlated landscapes, hence can actually *adapt better*. Large proteins live on landscapes with connected walks to higher optima than peptides do. And probably the linear structure of proteins ensures that the dominant interactions of any amino acid is with its few neighbours in the primary sequence. To a lesser extent, folding brings distant regions into contact, but any amino acid still interacts directly with only a few others. Thus, as the length of a protein becomes large, the number of strong interactions bearing on the contribution of any amino acid remains bounded and small. Proteins are thus probably well set up to adapt on correlated landscapes by their inherent nature.

What of genetic regulatory systems? What is low connectivity? Nothing other than the property that the one gene is directly regulated by only a few other genes or their products. But this is a restatement of *molecular specificity*. Enzymes and other liganding agents with high specificity will, in general, discriminate among a vast number of molecular variables and bind only a few close cognates. Selection for high specificity is also selection for low connectivity in a genomic regulatory system. Selection for low connectivity is, simultaneously, selection for systems which exhibit the globally ordered dynamics of homeostasis so reminiscent of actual cell types. In turn, it happens that selection for low connectivity and consequent homeostatic dynamical behaviour also yields genomic regulatory networks which adapt on highly correlated landscapes which avoid the complexity catastrophe!

Two points warrant attention. First, by selecting for large proteins, or high specificity, selection may be changing the adaptive landscapes of the kinds of entities which are evolving. Thus we do seem to confront the fact that selection can act on entities to achieve those with 'good' landscapes. This would not seem to require any group selection argument. This implies that we need to broaden evolution theory, not only to include the balancing forces of selection operating on systems with self-organized properties, but, further, we need to begin to understand how selection may operate to cull entities better able to adapt because they adapt on good landscapes.

Second, both in selection of large proteins and in genomic systems of low connectivity, the drive towards good primary function – proteins which are good enzymes and genomic systems with stable homeostatic dynamics – appears to bring with it (almost gratuitously) selection for entities which happen to adapt on 'good' landscapes. Is this fortuitous? Or is there some deeper reason hinting that complex systems which 'work well' also 'adapt well'?

Quite obviously, if we take the Kantian stance, and ask, 'what must organisms be, in order to evolve adaptively?', we become aware of how very much we have to learn. And if we look beyond evolution in biology to evolution in technological economies and society at large, we may well wonder whether general principles of adaptation in complex systems are to be found.

CONCLUSION

I have tried to suggest that evolution is a complex combinatorial optimization process on some form of rugged fitness landscape. This leads us to
realize that adaptation on such landscapes tends to become trapped on local
optima, or diffuse away from such optima if mutation rates are large enough
with respect to selective forces. We were led to wonder if there might be
quite general statistical features of rugged landscapes holding over many
combinatorial optimization problems. (A large literature on combinatorial
optimization exists, but not often from the evolutionary perspective. In
part this is based on simultaneous annealing: see Kirkpatrick, Gelatt and
Vechi, 1983).

Adaptive evolution confronts the profound opportunities afforded by the
self-organized properties of complex dynamical systems. I have here
broached only genetic regulatory networks, but similar issues appear to
arise in models of: the origin of life (Eigen and Schuster, 1979; Kauffman,
1986; Dyson, 1985; Rossler, 1974); connected metabolisms (Kauffman,
1986); and morphogenesis. (Over the last decade a very large number of
articles and books on morphogenesis have been published, based on A.
Turing's (1954) reaction–diffusion model of morphogenesis. Authors
include: Brian Goodwin, James Murray, Stuart Kauffman, Hans Meinhart,
Lionel Harrison, Lewis Wolpert, George Oster, Peter Bryant and Susan
Bryant. The articles are scattered in the following journals: *Science, Nature,
J. Theoret. Biol.* and *Developmental Biology*).

This surely implies that we must build a larger theory which marries
Darwin's idea of selection to the self-organized properties of the entities
that selection was privileged to operate upon. Here we confront the possible
limitations in the power of selection to avoid the typical properties of such
systems, hence potential universals in biology, and an escape from the
epistemological necessity of full reductionism. Beyond this, selection
appears able to fashion the entities it operates on to be ones with useful
landscapes. We thus need a still broader theory, encompassing this possibility, and then the probability that within such a 'selected' ensemble,
selection will still not be able to avoid the typical properties of that selected
ensemble, hence again an escape from necessary reductionism in our explanations. Only the surface has been touched. We need a true theory of
biological embracing self-organization, selection, and historical accident.

REFERENCES

Anderson, P.W. (1985) Spin glass Hamiltonians: a bridge between biology, stat-
istical mechanics and computer science. In: *Emerging Syntheses in Science.
Proceedings of the Founding Workshops of the Santa Fe Institute*. Santa Fe,
New Mexico.
Dyson F. (1985) *Origins of Life*. London: Cambridge University Press.

Eigen, M. (1985) Macromolecular evolution: dynamical ordering in sequence
    space. In: *Emerging Syntheses in Science. Proceedings of the Founding
    Workshops of the Santa Fe Institute.* Santa Fe, New Mexico.
Eigen, M. and Schuster, P. (1979) *The Hypercycle.* New York, Heidelberg:
    Springer-Verlag.
Ewens, W.J. (1979) *Mathematical Population Genetics.* New York, Heidelberg:
    Springer-Verlag.
Gillespie, J.H. (1974) *Theoret. Population Biology 44,* 167.
Hopfield, J.J. (1982) *Proc. Natl. Acad. Sci. 79,* 254.
Kauffman, S. (1969) Metabolic stability and epigenesis in randomly constructed
    genetic nets. *J. Theoret. Biol. 22,* 437.
——(1974) The large scale structure and dynamics of gene control circuits: an
    ensemble approach. *J. Theoret. Biol. 44,* 167.
——(1984) Emergent properties in random complex automata. *Physica 10D,*
    145.
——(1986) Autocatalytic sets of proteins. *J. Theoret. Biol. 119,* 1.
——(1989) *Origins of Order: Self-organization and Selection in Evolution.* Oxford
    University Press.
Kauffman, S. and Levin, S. (1987) Towards a general theory of adaptive walks
    on rugged landscapes. *J. Theoret. Biol. 128,* 11–45.
Kauffman, S. and Smith, R.G. (1986) Adaptive automata based on Darwinian
    selection. In: Farmer *et al.* (eds) *Evolution, Games and Learning. Models for
    Adaptation in Machines and Nature.* Amsterdam: North-Holland Press.
Kirkpatrick, S., Gelatt, C.D. Jr and Vechi, M.P. (1983) Optimization by sim-
    ulated annealing. *Science 220,* 671.
Lin, S. and Kernighan, B.W. (1973) An effective heuristic algorithm for the
    travelling salesman problem. *Oper. Res. 21,* 498.
Rossler, O. (1974) Chemical automata in homogeneous and reaction diffusion
    kinetics. Notes in *Biomath. B4,* 399.
Smith, John Maynard (1970) Natural selection and the concept of a protein
    space. *Nature 225,* 563.
Wright, S. (1932) The roles of mutation, inbreeding, crossbreeding, and selec-
    tion in evolution. In: *Proceedings of the Sixth International Congress on
    Genetics I,* 356.

7. Evolution and the generative order

B.C. GOODWIN

The biological process that is known in contemporary biology as development used to be referred to as generation, implicating creative origins as in
genesis. We find this use of the term in, for example, William Harvey's
(1651) studies of the developing chick, *De Generatione Animalium*, a use
which persisted into the 19th century. However, by this time alternative
expressions were becoming more common, and towards the end of the last
century Weismann (1883) was referring to 'generation' as 'growth', while
Roux (1884) adopted the term development (*Entwicklung*) to define his
embryological research programme. An explicit connection between
genesis and heredity came about with Bateson's (1906) introduction of the
term 'genetics' to describe the study of Mendelian inheritance. He emphasized the particulate atomic nature of Mendelian units (later genes) and
the quantized quality of inherited phenotypic differences. At the same time,
Bateson (1894) recognized the role of developmental processes in generating the discontinuities that are one of the most striking features of biological
form, such as asymmetries (left- or right- handed structures – e.g. limbs)
and periodic or meristic series of structures (e.g. petals, segments, digits,
etc.). Thus he saw both genes and generation as components in the process
which generates what he called, following Galton, 'Positions of Organic
Stability'. These are the stable morphological states that result from morphogenesis.

This balance has not been maintained throughout the present century,
the burden of generative origins tending to fall more and more on the genes
rather than on morphogenetic mechanisms. Thus, rather then following
Weismann's (1883) injunction to 'trace heredity back to growth' (i.e. development or reproduction), there has been a progessive tendency to trace
development back to heredity. Parodoxically, Weismann himself contributed substantially to this shift of focus by his germ plasm/somatoplasm
dualism, the former being endowed with 'the power of developing into a
complex organism' (Weismann, 1885). Despite this progressive shift of
emphasis during the course of this century, there have been those who
espoused and developed the integrative position defined by Bateson,
noteable examples being Needham (1932) and Waddington (1975).

Integrating evolution and development, or genes and generation, into a coherent conceptual framework has now become a compelling, challenging task in biology. Both are concerned with different aspects of biological origins, with genesis: evolution addresses the origin of taxa; development the origin of structure and ordered complexity in organisms. Two different types of explanation have emerged for these aspects of generative order. In evolution, the dominant mode of analysis, following Darwin (1859), is genealogical or historical. The origin of species is taken to be a result of historical processes, hereditary variation and natural selection. The variety of developmental strategies that generate the diversity of present species is therefore understood to be a consequence of the survival of fitter ontogenies producing and reproducing fitter organisms. Differential survival is regarded as the causal origin of the discrete spectrum of organismic diversity that underlies taxonomy. This perspective on the relationship between development and evolution gives rise to a particular research programme: the identification of appropriate units of selection, the reconstruction of plausible selective scenarios, and the search for evidence that persisting ontogenies are indeed the fitter variants. The goal of this programme is the realization of Darwin's vision: 'Our classifications will come to be, as far as they can be so made, genealogies; and will then truly give what may be called the plan of creation'.

In contrast with this historical context of explanation, developmental studies on the epigenetic origins of complex organismic forms seek explanations of the type used in the so-called exact sciences. These attempt to describe necessary and sufficient conditions for the production of a particular state or structure, whereas historical explanations are necessarily restricted to the description of some necessary conditions only. The point at issue here is intelligibility: how are we to comprehend the generative origins of spaces as distinctive forms and life-histories? Research on morphogenesis is based on the ontogenies of existing species, with the goal of describing the molecular, cellular, physical and chemical processes that are necessary and sufficient to describe (predict) the generation of particular structures. Although precise modelling of developing organisms is still in a rudimentary state, the objectives of developmental biology have been explicit since Roux enunciated the principles of *Entwicklungsmechanik* over a century ago.

These contrasting explanatory styles of evolutionary and developmental studies pose interesting problems for a resolution of the two areas of biology concerned with questions of origins. Looked at formally, there are serious difficulties associated with the evolutionary research programme. Historical reconstruction of differential survival among competing lineages based upon present stable states (persisting ontogenies) is not a well-posed problem, as one says in mathematics, because there is always an unlimited, undefined set of possible solutions. The essential difficulty is that from a knowledge of the stable states of a process (i.e. its attractors, which in the case of evolution are the life cycles of existing species), one cannot deduce the dynamics that generates the trajectories leading to these states. One needs to describe the dynamics first and then deduce the structure and distribution of the attractors.

This procedure is actually followed in population genetics whenever a specific model is studied. A particular dynamic is assumed, usually gene frequency changes in a population dependent upon the birth and death process of organisms, immigration and emigration, etc., which together with selection coefficients result in equations whose stable states or attractors define the gene frequencies of a population. However, this approach clearly ignores development, ontogeny. It cannot be used to construct morphological classifications because it does not involve morphogenesis. Therefore the realization of Darwin's taxonomic ambitions is left to other methods of historical reconstruction which suffer from the difficulty mentioned above: there is always an unlimited set of possible trajectories leading to particular morphological states if all one has to go on is the existence of these states and some genealogical relationships between them. These are not sufficient to describe the generative dynamics of biological forms.

The question then arises whether it is possible to understand the generation of stable ontogenies using the same analytical procedures that are used in population genetics (procedures that are common to all exact analyses, which describe necessary and sufficient conditions for a process and so result in well defined solutions). This requires that the dynamics of development be described (which clearly includes gene activity), followed by a study of the resulting stable morphological states, Bateson's 'Positions of Organic Stability'. Just as population genetics is based upon what are assumed to be universal properties of the hereditary process, so the study of ontogenies is based upon what are assumed to be universal properties of the developmental process, which make possible the emergence of the diversity of organismic forms characteristic of different life histories (species). If such an analysis can be realized, then it may be possible to understand taxonomy in terms of the logical, relational order in this generated set, while evolution is the time-dependent exploration of its multifarious dynamic potential. Developmental dynamics then becomes the generative basis of taxa and their temporal emergence in evolution, while 'natural selection' refers to the dynamic stability of the generated attractors. Natural selection is not, evidently, a causal dynamic principle; it is a description of stability.

This research perspective belongs within a well-defined tradition of biological study whose morphological roots extend back to Goethe (see Brady, 1986) and Geoffroy St. Hilaire and includes the work of Roux and Driesch (cf. Russell, 1916), D'Arcy Thompson (1942), Needham (1932), and, most significantly for this volume, Waddington (1957). Within the past two decades or so, this orientation has been identified as part of the twentieth-century structuralist movement (Piaget, 1970; Thom, 1972; Webster and Goodwin, 1982; Lambert and Hughes, 1984). This is founded upon a tradition of rationalism which is critical of historical, functionalist theories such as neo-Darwinism because of their failure to address issues of generative origins ('dynamics') and their preoccupation with stability as a consequence of historical particulars, which remain unintelligible because they simply happen. A major preoccupation of structuralist biology can be defined as the problem of biological form (Webster and Goodwin, 1982). In this essay I shall explore this problem in relation to specific aspects of development, with a focus on morphogenetic principles as the generators of organismic form.

THE EVOLUTION OF GASTRULATION

A basic conundrum in biology is why organism ever become more complex than the unicellulars that populated the Precambrian oceans, whose 'success' can be measured by the teeming millions of micro-organisms that continue to pursue similar life histories to those of their ancient progenitors. One of the most significant steps in the emergence of greater organismic complexity was the evolutionary origin of gastrulation. The transformation of a hollow ball of cells into a multilayered structure, with the consequent combinatorial potential for reciprocal inductive interactions leading to diverse patterns of cell differentiation, stands out as a major event in the evolution of the metazoa. How are we to understand the origin of this process? We have on offer two rather different types of answer. Consider first the analysis presented by Leo Buss (1987) in his book *The Evolution of Individuality*. I shall make extensive use of the arguments in this intriguing volume because of the clarity with which functionalist explanations are developed.

Buss's primary objective is to reintegrate development with evolution, something that the synthetic theory side-stepped by adopting Weismann's dualistic germ plasm/somatoplasm description of organisms. Weismann's idealist (indeed, vitalist) proposal that the germ plasm has the organizing properties required to produce an organism from the largely passive somatoplasm has been developed into the equally vitalist notion that there is a part of the organism, the so-called genetic programme, that is responsible for the developmental process. This is vitalist because it ascribes to the genetic programme organizing properties (the power of generating an organism of a particular form) which it does not have (Stent, 1982; Goodwin, 1985). It confuses the valid concept that hereditary phenotypic *differences* between organisms are correlated with particular genetic differences, with the erroneous deduction that development itself can be explained by the action of genes. This is a mistake that geneticists who are brought up on the clarity of Haldane's definitions rarely make (cf. Maynard Smith, 1965).

However, it is not this error with which Buss is concerned. He points out that most organisms, over most of evolutionary time, failed to conform to Weismann's germ plasm-somatoplasm dualism, because the reproductive organs are generated from the 'somatoplasm' during ontogenesis. The hereditary unit on which selection acts is therefore not, in general, the whole individual organism (hereditarily continuous with the germ plasm in Weismann's theory), but some part of it that is generated during development. Buss concludes that: 'A synthetic theory came into being that was and is at variance with known developmental patterns' (p. 4). His book is concerned to rectify this situation. However, he assumes that: 'The synthetic theory cannot be incorrect; it can only be incomplete' (p. 25). Buss therefore undertakes to show how natural selection, acting on inherited variants, can explain the major phenomena of development. Since most reproductive strategies do not obey Weismann, most organisms are not themselves individual units of inheritance and so it is necessary to define the actual units of selection in these ontogenies. According to Buss, they are cells. His exploration of the evolutionary origin of gastrulation then proceeds as follows.

NATURAL SELECTION AND THE GASTRULA

Buss starts with the observation that the cells of a metazoan can be either ciliated or they can divide, not both. The reason is that mitotic spindles and undulipodia (cilia or flagella) both require microtubule organizing centres, and in cells such as those of the metazoa with one such centre per cell there is a conflict such that one or other structure is possible, not both. Imagine, then, protists with these characteristics aggregated to form a hollow ball of cells. Buss suggests that this ball faces a problem: it needs to be ciliated so that it can undergo dispersion and act as an efficient propagator; and it needs to continue cell division if it is to develop into a more complex organism. But these states are incompatible. How does the ball 'solve the problem'? It forms a gastrula. The cells on the surface remain ciliated, while some move to the interior where they lose their cilia and so can divide. 'Animal gastrulation is the metazoan solution to the requirement of simultaneous development and movement (p. 44).

What kind of explanation is this, and what are its predictions? Buss says that 'the demand for simultaneous ciliation and continued development gave rise to gastrulation' (p. 65). This 'solution to a problem' scenario, which is symptomatic of the cognitive style of explanations by natural selection, is frankly and explicitly teleological and needs to be recast to conform to accepted notions of causality. This requires that the 'correct solution', the ciliated gastrula, has a selective advantage over other forms and so will increase as a result of differential survival. This is what is meant by solving the problem. A simple prediction is that the forms that failed to find the solution should have been eliminated. Sponges failed: they form an unciliated ball which then gastrulates, but they seem to be surviving well. Colonial organisms failed: they form hollow ciliated balls that do not gastrulate or develop further, but they, too, continue to survive. Many protists are not constrained by the either/or limitation: they can be ciliated and divide at the same time. Here Buss makes an interesting remark: in ciliates 'macronucleus division is a complex and inefficient process'. Given the success of the ciliates, it is not clear what objective criterion is being used here for efficiency.

Looked at from the perspective of explanation in the exact sciences, these arguments are exceedingly bizarre. There are three major difficulties with 'explanations' by natural selection. First, the selective scenario proposed is untestable because the conditions that prevailed cannot be reconstructed. Second, the simple predictions that appear to follow but are violated can all be explained away by inventing other selective scenarios for surviving forms that did not solve the problem, and so the hypothetical explanation is not falsifiable in these terms. Third, and most seriously, no account is given of what causal process may have actually generated the gastrula. Yet this is the only type of explanation that would be accepted in an exact science. And this is what developmental biology is about. Roux proposed *Entwicklungsmechanik* as the research programme whose objective is to understand the generative causes of epigenesis, partly in reaction to the claims of nineteenth-century Darwinists to have discovered the real causes of ontogenesis in natural selection (Russell, 1916; Cassirer, 1950). If there is to be a unification of development and evolution, it must be in terms of an analysis that attempts to articulate the evolution of developmental processes as consequences of clearly-defined dynamic causes. Natural selection itself explains nothing about generative dynamics. Differential survival is a consequence of this dynamics, not a cause. Furthermore, the 'units of selection' are also consequences of the dynamics, the (relatively) stable attractors in the evolutionary process (stable life histories). They cannot be defined initially. In order to understand the evolutionary origins of the gastrula, we must proceed with the analysis in a different way. The basic structure of this analysis was provided several years ago by Willmer (1960). Oddly enough, Buss gives no reference to this very interesting work.

THE DYNAMIC BASIS OF GASTRULATION

In the course of his cytological studies, Willmer observed that a number of species of protists have the interesting property that they undergo dramatic, reversible changes of state, from flagellate to amoeboid, depending on whether they are in a low or high ionic strength medium, respectively. Suppose that protists with these properties aggregated to form a hollow ball of cells in the Precambrian oceans, as assumed by Buss. Such a form is a minimal energy configuration for such an aggregate, making reasonable assumptions about cell affinities, so this assumption is dynamically well-founded. Due to the osmoregulatory properties of cells, which pump ions across their boundaries to maintain physiological ratios of ions, the confined internal space of the hollow ball will have a higher ionic strength than the external medium. These observations therefore suggest that cells on the outside of the ball would tend to be flagellated or ciliated, while any cells that moved into the interior as a result of the balance of forces between cell–cell adhesion, compression of the ball, and spontaneous movement, would tend to become amoeboid, in which state they can divide. This is the essence of Willmer's proposals, which he developed in relation to the different patterns of gastrulation that occur in different species.

Hence we have a perfectly acceptable hypothesis about the origins of the gastrula in terms of known properties of protists and the balance of forces acting within and on a spherical cell aggregate. Furthermore, a whole series of experiments are possible to explore this hypothesis further. One process that is not understood is the flagellate–amoeboid transition under a change of ionic strength. Some might suggest that this property is a consequence of natural selection. But this amounts to the suggestion that there must be come natural causal explantion of the phenomenon, and that it must be consistent with the dynamic stability of certain protist life-cycles in the context of particular environments. This adds nothing to a research programme attempting to understand generative causes. What one wants to know is the balance of forces acting on cells, the effects of ionic strength on the cytoskeleton and the different states of spatial order that this dynamic structure can assume under different conditions of Na^+, K^+, Ca^{2+}, H^+, Cl^-, etc., including the influence of ionic fluxes and electrical current flows that could break symmetries and generate spatial patterns in different ways. Such studies, being pursued in a variety of laboratories, define the basic dynamics of cell organization that provide some of the universal generative principles for evolutionary events.

THE EVOLUTION OF GENERIC FORMS

Within this perspective, a distinctly different dimension is added to the functionalist selective scenario. What becomes evident from Willmer's proposals about the origin of the gastrula is that this state of spatial organization, the ball of ciliated cells with dividing amoeboid cells inside, may be a robust and inevitable consequence of the forces acting on and within a cellular aggregate. But it is not the only one. Other states may be equally robust and inevitable, given other parameter values; for example, the non-ciliated gastrula of the sponge, or the ciliated non-gastrula of colonial organisms, may be equally natural forms. These are simply different possible states, different attractors for different parameter values in a self-organizing dynamic system. Each of these forms must, of course, be able to survive and reproduce in order to persist. This is what is meant by dynamic stability. The problem of reintegrating development and evolution is that of describing, on the basis of all relevent empirical and theoretical information, the dynamics of whatever process one wishes to study, to model it and show what the attractors are, and to check these predictions experimentally. What then becomes a perfectly reasonable proposal is that persisting states of biological order (ontogenies, life-cycles) are the high-probability, robust, or natural states of the developmental–evolutionary dynamic; i.e., they are the *generic states* of this process in the sense that they are typical, what one expects to find as natural forms (Webster, 1984), like waves on liquids. These generic states or forms may be highly discrete, well-defined attractors (Odell *et al.*, 1981) whose distribution in the space of developmental trajectories defines the set of possible forms and hence the taxonomy of their generative relationships (Goodwin, 1989; Webster, 1989). Evolution is then the time-dependent exploration of this set of possibilities under internal (genetic) and external (environmental) parametric variation. Some states are metastable and lead on to the variety of metazoan ontogenies; others, such as ciliated colonial forms that do not gastrulate, are more limited in their developmental potential. What one sees revealed in evolution is, then, the generic states of this developmental dynamic. The same proposal has emerged from a quite different approach to this problem by Kauffman and Levin (1987) (see also Kauffman, this volume), who studied the behaviour of organisms as complex epistatic genetic networks. Their conclusion was that the states occupied by such networks are overwhelmingly generic.

STRUCTURALIST BIOLOGY

This perspective results in the research programme of structuralist biology (Goodwin *et al.*, 1989). Its objective is to define the generative principles that underlie the production and reproduction of organisms, based upon the properties of present ontogenies. Knowing the dynamics, one can classify the attractors, which are the 'units of selection'. Guessing at the nature of these units on the basis of elementary properties, such as the capacity to replicate or reproduce, can lead one astray. We see this in Dawkins' (1986) concept of the replicator as the essence of life, the unit of selection, which falls into Weismann's trap for two reasons. First, the DNA replicator is not autonomous; in all non-parasitic organisms the complete and accurate replication of the hereditary material is a property of a cell, not a molecule. Second, this replicator does not contain the information to make an organism (Stent 1982, Goodwin 1985).

This difficulty also becomes evident in Buss' (1987) concept of the cell as the unit of selection in the evolution of ontogenies. If this is so, then dividing lineages within the metazoan gastrula are in competition with one another to leave the maximum number of progeny. The evolution of higher forms is then dependent upon the chance emergence of variant cell lineages within a developing organism which 'further their own replication rate by restraining or directing the activities of neighbouring cells' (p. 78). Inductive interactions between cells therefore have a specific character: the induced lineage undergoes a reduction in replication rate relative to the inducing lineage. Certain aspects of this perspective can be studied experimentally, others not. As Buss himself admits: 'A precise phylogenetic reconstruction of each interaction between variant cell lineages . . . is quite impossible at present' (p. 90). His hope that 'the "fossil record" writ in the genome may, eventually, provide such a reconstruction' (p. 90) is clearly not a sound basis for a research programme. However, other aspects of his proposals can be experimentally pursued.

First, is it true that induced cell lineages suffer a reduction in replication rate? Unless 'induction' is unreasonably restricted, there are many examples in which derivatives of one cell lineage actually induce cells from another lineage to replicate faster. An instance is the apical ectodermal ridge of the chick (Wolpert, 1981) or the amphibian (Fallon and Crosby, 1977) limb bud, which maintains conditions for rapid proliferation of the underlying mesenchyme cells from which the bone, muscle, and other mesodermal derivatives of the limb arise. Second, can metazoan development be understood in terms of the interactions of cell lineages? Whereas this is *largely* true for certain species, the nematode *Caenorhabditis elegans* being probably the best-known example, it is not generally the case. Even in *Drosophila*, where compartments tend to restrict lineages to defined components of adult morphology, a particular part of the adult can be made of cells from different lineages. For example, some cellular components of the anterior bristles of tarsal segments of the limbs can come from the posterior limb compartment (Morata and Lawrence, 1982); while in mammals and amphibians, the general rule is that any morphological structure is made up of cells from a variety of lineages. Morphogenesis is not a respector of cell provenance. This is because there are many levels of dynamic order above cells in ontogenetic process. Cell sheets behave as viscoelastic units; epiboly is not understandable in terms of the behaviour of single cells; the dynamics of pattern formation in plant meristems (Green, 1987) or of pigment patterns in animals (Murray, 1981) requires a description of fields of particular types of force, not replicating cell lineages, though the latter may well be components of such fields.

It is of interest to note that Buss reinterprets the branching valleys of

Waddington's (1957) epigenetic landscape as variant cell lineages in competition with others. The usual perspective is to understand the diverging valleys as bifurcations in an epigenetic dynamic whose spatial elements are morphogenetic fields generating specific morphologies such as wings or halteres in *Drosophila*. These are alternative attractors into which imaginal discs can develop, between which they can switch as a result of either genetic (Lewis, 1975) or environmental (e.g. ether; see Ho *et al*., 1983) perturbations. Similarly, balls of cells can develop into different morphological states, a ciliated gastrula, an unciliated gastrula, or a ciliated colonial form, depending upon genetic and environmental parameters. This perspective attemps to define the dynamics of epigenesis in terms of the actual forces operating in the system, and to deduce its dynamically stable states as space–time solutions of self-organizing processes, all of which are transformations of one another (Goodwin, 1988; Webster, 1984). Some of these will have the characteristics of limit cycles in a certain sub-space, so that life cycles occur; others will not. The 'units of selection' are simply these dynamically stable life cycles, and natural selection says that these will be found in relative abundance reflecting the dynamics of their interactions, a statement about the (quasi-) steady states of the evolutionary process.

CONCLUSION

The integration of development and evolution requires that the real generative processes underlying stable life cycles be understood. One of the objectives of a structuralist analysis in biology is to classify the dynamically stable states of these processes based upon the general properties of ontogenetic processes, taking account of all levels of organization involved. Once this has been achieved, it will be possible to attempt historical reconstructions of actual phylogenetic sequences if these answer what are perceived to be significant questions. But it may well emerge that, if stable life cycles are in large measure the generic states of ontogenetic dynamics, then most of the major biological questions about the relationship of different taxa to one another, which is what Darwinian evolution was intended to answer, will receive explanations in generative terms which are independent of history. Biology will then have become an exact science.

REFERENCES

Bateson, W. (1894) *Materials for the Study of Variation*. Cambridge University Press.
——(1906) *The Progress of Genetic Research*. Inaugural Address Royal Horticultural Society. Report.

Brady, R.H. (1986) Form and cause in Goethe's morphology.

Buss, L.W. (1987) *The evolution of Individuality*. Princeton University Press.

Cassirer, E. (1950) *The Problem of Knowledge*. New Haven: Yale University Press.

D'Arcy Thompson (1917) *On Growth and Form*. Cambridge University Press.

Darwin, C. (1859) *The Origin of Species*. First edition. Harmondsworth: Penguin.

Dawkins, R. (1986) The Blind Watchmaker. New York: W.W. Norton & Co.

Fallon, J.F. and Crosby, G.M. (1977) Polarizing zone activity in limb buds of amniotes. In: Ede, D.A., Hinchcliffe, J.K. and Balls, M. (eds), *Vertebrate Limb and Somite Morphogenesis*, pp. 55–71. Cambridge University Press.

Goodwin, B.C. (1985) What are the causes of morphogenesis? *Bio Essays 3*, 32–35.

Goodwin, B.C. (1988) Morphogenesis and heredity. In: *Evolutionary Processes and Metaphors* (eds. M-W. Ho and S. Fox), pp. 145–62.

——(1989) The evolution of generic forms. In: *Organizational Constraints on the Dynamics of Evolution*. (G. Vida and J. Maynard Smith, eds). Manchester University Press (in press).

Goodwin, B.C. Sibatani, A. and Webster, G.C. (eds) (1989) *Dynamic Structures in Biology*. Proceedings of the first International Conference on Structuralist Biology, Osaka, Japan. Edinburgh University Press.

Green, P.B. (1982) Biophysics of the extension and initiation of plant organs. In *Developmental Order: its Origin and Regulation*, pp. 485–509. New York: Alan R. Liss.

Ho, M-W., Tucker C., Keeley, D. and Saunders, P.T. (1983) Effects of successive generations of ether treatment on penetrance and expression of the *bithorax* phenocopy in *Drosophila melanogaster*. *J. Exp. Zool. 225*, 357–68.

Kauffman, S.A. and Levin, S. (1987) Towards a general theory of adaptive walks on rugged landscapes. *J. Theoret. Biol. 128*, 11–45.

Lambert, D.M. and Hughes, T. Misery of functionalism. Biological function: a misleading concept. *Rivista di Biologia 77*, 477–501

Lewis, E.B. (1963) Genes and developmental pathways. *Amer. Zool. 3*, 33–56.

Maynard Smith, J. (1965) *Evolution*. London: Penguin.

Morata and Lawrence (1982)

Murray, J.D. (1981) On pattern formation for lepidopteran wing patterns and mammalian coat markings. *Phil. Trans. Roy. Soc. B. 295*, 473–96.

Needham, J. (1932) *Order and Life*. Yale University Press.

Odell, G., Oster, G.F., Burnside, B. and Alberch, P. (1981) The mechanical basis of morphogenesis. *Devel. Biol. 85*, 446–62.

Piaget, J. (1980) *Adaptation and Intelligence: Organic Selection and Phenocopy*. University fo Chicago Prress.

Roux, W. (1894)

Russell, E.S. (1916) *Form and Function*. London: Murray.

Stent, G. (1982) What is a program? In: *Evolution and Development*, pp. 111–13. Dahlem Workshop Report 22. Berlin: Springer-Verlag.

Thom, R. (1972) *Stabilité Structurelle et Morphogénèse*. Reading: Benjamin.

Waddington, C.H. (1957) *The Strategy of the Genes*. London: Allen & Unwin.

Webster, G.C. (1984) The relations of natural forms. In: M-W. Ho and P.T. Saunders (eds.) *Beyond Neo-Darwinism*, London: Academic Press.

——(1989) Structuralism and Darwinism: concepts for the study of form. In: (eds.) Goodwin, B.C., A. Sibatani and Webster, G.C. *Dynamic Structures in*

Biology Proceedings of the First International Conference on Structuralist Biology, Osaka, Japan. Edinburgh University Press (in press).

Webster, G. and Goodwin, B.C. (1982) The origin of species: a structuralist approach. *Journal of Social and Biological Structures* 5, 15–47.

Weismann, A. (1883) Reprinted in Hall, T.S. (ed.) (1964) *A Source Book in Animal Biology*. New York: Hafner.

——(1885) Reprinted in Moore, J.A. (ed.) (1972). *Readings in Heredity and Development*. New York: Oxford University Press.

Willmer, E.N. (1960) *Cytology and Evolution*. New York: Academic Press.

Wolpert, L. (1981) Pattern formation in limb morphogenesis. In: Sauer, H.W. (ed.) *Progress in Developmental Biology*, pp. 141–52. Stuttgart: Fisher.

8. The Physics of Complex Organisation

P.C.W. DAVIES

GOODBYE TO VITALISM AND REDUCTIONISM

One of the great outstanding challenges to science is the reconciliation of physics and biology. Biological organisms have always excercised a strange fascination for the physicist, because they (that is, the organisms) do such clever things. However, physicists have traditionally balked at the study of the living phase of matter because organic processes seem so impenetrably complex.

Until recently, complexity was seen as mere complication. Today, largely thanks to the electronic computer, complex systems have been studied in great generality, and there is a growing recognition that complex processes can display certain systematic properties and regularities. Indeed, many researchers speculate that there are definite principles of complexity that enjoy a universal, or at least a quasi-universal, status. It then becomes an important matter to accommodate these principles within the underlying laws of physics.

The most conspicuous property of a living organism is not complexity per se, but organization. The complex processes that occur are of a cooperative nature, thus endowing the organism with a coherent identity as a whole. It is this impression that the organism is somehow *harnessing* the blind and purposeless forces of physics and arranging them, to use Monod's words, as part of a plan or project, that seems so baffling. What is the source of this 'purpose without purpose'?

Two great conceptual obstacles have traditionally plagued attempts to solve this riddle: reductionism and vitalism. The former regards complex systems as the indifferent conjunction of many components, a peaceful coexistence or superposition of elementary units. This image of complexity as 'mere complication' is a legacy of the age of linear physics, which began in earnest with Newton and endured until well into this century. In a linear system, such as electromagnetism, reductionism is a powerful methodology. Indeed, much of eighteenth and nineteenth century mathematics, such as Fourier analysis and the techniques for the solution of differential equations, were directed to the study of linear physical systems. The very word 'analysis' implies that one can understand a complicated system by elucidating the qualities of its components in isolation, and then superimposing them, knowing that the whole is merely the sum of its parts.

Because living organisms are nonlinear (for a start!) any attempts to explain their qualities as wholes by the use of analysis is doomed to failure. There are many simpler, moderately well-understood physical systems such as assemblages of self-gravitating bodies, solitons and turbulent fluids, where the failure of reductionistic analysis is manifest. This is not to denigrate the very important work of molecular biologists who carefully unravel the chemical processes that take place in living systems, only to recognize that understanding such processes will not *alone* explain the collective, organisational properties of the organism.

Before it became appreciated that qualitatively new features can, in a nonlinear system, emerge from the conjunction of component parts ('the whole is greater than the sum of its parts') some biologists despaired of ever explaining biological organization in terms of ordinary physical forces. This led to the blind alley of vitalism: the gratuitous introduction of hypothetical 'vital forces' to augment those known to physics. Vitalism was a response to reductionism. When the whole is regarded as merely the sum of its parts, an animate whole cannot be composed of inanimate parts.

The most cogent argument against vitalism is the existence of self-organizing processes in nonbiological science. Many nonlinear systems have been discovered to possess the remarkable ability to leap spontaneously from relatively featureless states to those involving complex cooperative behaviour. In thermodynamics it is found that certain physical and chemical systems, when driven far from equilibrium, arrange themselves into so-called dissipative structures (Nicolis and Prigogine 1977). A much-cited example concerns the Bénard instability, in which a fluid is heated carefully from below. At a critical temperature gradient, the initially homogeneous fluid spontaneously organizes itself into a pattern of convective flow. The pattern can be highly distinctive, displaying well-ordered hexagonal cells (Tritton 1988).

The striking point about this simple phenomenon is that although each molecule of the fluid is merely pushed or pulled by its immediate neighbours, countless trillions of molecules conspire to produce long-range order without any help from outside. No unseen hand shapes and supervises the pattern of flow, yet no individual molecule is aware of the activity of any but its closest companions.

Similar behaviour is found in the laser, wherein myriads of excited atoms spontaneously cooperate to emit their light exactly in phase, producing a perfectly sculptured waveform, rather than the incoherent jumble of wavelets characteristic of an ordinary lamp. Certain chemical mixtures, forced far from equilibrium, grow elaborate structures or pulsate rythmically. These and many other examples of self-organization have been studied by Ilya Prigogine and his school. (For a popular review see Prigogine and Stengers 1984). Images of self-organising behaviour have also been captured in mathematical models of complexity, such as cellular automata and network theory (Farmer *et al.* 1984, 1986).

Only the most die-hard vitalist would invoke occult forces to account for convection cells or laser light. Mathematically we can now see how non-linearity in far-from equilibrium systems can induce matter to 'transcend the clod-like nature it would manifest at equilibrium, and behave instead in dramatic and unforeseen ways, molding itself for example into thunder-storms, people and umbrellas' (Bennett 1986).

In this post-reductionist era of physics, one looks not to the individual molecules for an explanation of large-scale self-organization, nor even to the action of one molecule upon its neighbour. Instead attention is focussed on *global* issues, such as boundary conditions, constraints, distance from equilibrium and so on. True, the properties of individual molecules and their interactions are necessary ingredients in providing a full explanation, but they are not *sufficient*; the key concepts are left out. I believe that biological organisms differ only in the degree of their organisational capa-bilities from the simpler foregoing examples. Therefore it would be rash to conclude that in the biological case molecules alone explain all.

Although many biologists seem reluctant to accept this, physicists gener-ally seem to have taken it on board. For example, in their discussion of chaotic systems (see below) Crutchfield *et al.* (1986) write:

Chaos brings a new challenge to the reductionist view that a system can be understood by breaking it down and studying each piece . . . a system can have complicated behaviour that emerges as a consequence of simple, nonlinear interaction of only a few components. The problem is becoming acute in a wide range of scientific disciplines, from describing microscopic physics to modelling macroscopic behaviour of biological systems . . . For example, even with a complete map of the nervous system of a simple organism . . . the organism's behaviour cannot be deduced . . . The interaction of components on one scale can lead to complex global behaviour on a larger scale that in general cannot be deduced from knowledge of the individual components.

ORGANISING PRINCIPLES: BACKDOOR VITALISM?

There is an age-old dialogue between reductionists and their opponents – holists. The holist believes in *emergence*; that is, at each level of complexity in the hierarchy of physical systems new qualities emerge that are not only absent, but are clearly meaningless at lower levels. For example, the concept 'electrical resistance' cannot be applied to individual electrons, but it is perfectly meaningful in the context of electric circuit theory. There is even a law – Ohm's law – about it. Likewise in biology, concepts like natural selection, adaptation, conscious awareness and even the quality of being alive find no application to the individual atoms on one's body.

Now the reductionist does not deny the utility of these higher level concepts, but she argues that they are not *fundamental* in their own right. They can, she alleges, be *reduced* to the properties of the elementary particles and fields that constitute all matter. The fact that in almost all cases nobody has actually demonstrated this reduction is no discouragement. *In principle* such a reduction could be effected, she claims. Thus, to carry the argument to its logical extreme, one could account for, say, the migratory habits of birds or my thoughts about this article entirely in terms of the superstring Lagrangian (or whatever is the fashionable model for the ultimate building blocks of matter (Davies and Brown 1988)).

Now I regard this final statement as so self-evidently absurd that I suspect that nobody would seriously support it. Yet if the statement is false it is clear that we must add *something else* to the superstring Lagrangian (or whatever) to explain bird migration and thinking. But what else? Where in the hierarchy from strings to quarks and leptons, to nuclei to atoms to molecules, to DNA and proteins, to cells to birds and people do we encounter 'something else'?

Before the reader rushes to the conclusion that I am trying to slip in vitalism by the back door let me discuss a concrete example. When I play Space Invaders on my computer, the shapes on the screen move about according to certain simple dynamical rules. The physical phenomena that produce these moving shapes are, of course, subject to the laws of physics. So does that mean that one could, in principle, deduce the dynamical 'laws' of Space Invaders from an examination of the laws of physics?

Clearly not. The 'secret' of a computer (cf the 'secret' of life) is not in the laws of physics as such (though those are necessary), but in the wiring; more specifically, in the topology of the circuitry. Then one has to go into the software aspect, and determine how that is encoded in the machine. These features, which are in addition to the laws of physics but contain the key to how Space Invader shapes move, are called *constraints*.

In almost every branch of physics, with the possible exception of elementary particle physics, constraints play a crucial role. Even a problem as simple as the motion of the humble pendulum depends both on Newton's laws of dynamics and on constraints – such as the length of the string. Attempts to derive 'secondary' laws of physics, such as Ohm's law, from the underlying 'primary' laws, such as the motion of electrons, always require the specification of constraints in addition to the primary laws. Notice that it is our freedom to choose the constraints that enables us to impose systematic rules such as Space Invader dynamics without in any way threatening the laws of physics. There are two levels: the underlying laws of physics that determine how the electrons, photons, etc, move about inside the machine, and the higher-level 'laws' having to do with the *organisational* aspects of that vast assemblage of electrons, photons, etc. There is complete compatability between these two sets of laws.

What is good for computing machines is good for brains. Here we replace the visual display by the realm of thoughts. Are the 'laws of thought' reducible to the laws of physics? Marvin Minsky has addressed this (Minsky 1987):

> Many scientists look on chemistry and physics as ideal models of what psychology should be like. After all, the atoms in the brain are subject to the same all-inclusive laws that govern every other form of matter. Then can we also explain what our brains actually do entirely in terms of those same basic principles? The answer is no, simply because even if we understand how each of our billions of brain cells work separately, this would not tell us how the brain works as an agency. The 'laws of thought' depend not only upon the properties of those brain cells, but also on how they are connected. And these connections are established not by the basic, 'general' laws of physics, but by the particular arrangements of the millions of bits of information in our inherited genes. To be sure, 'general' laws apply to everything. But, for that very reason, they can rarely explain anything in particular.

Let me now deal with a possible objection to all this. An out-and-out reductionist might argue that all constrained systems are constrained *by* something, and those somethings will also be subject to the laws of physics. Thus might not an application of those laws enable us to *derive* the all-important constraints, rather than having to impose them *in addition* to the laws of physics? This objection can be sustained if one is willing to revert to a Newtonian–Laplacian closed mechanistic universe, wherein all future states of matter are completely determined at every level of detail by the state at any given moment. In such a universe everything that ever happens, such as somebody deciding to build a computer and somebody else inventing Space Invaders, or the precise wiring of my brain, is fixed since time immemorial. Now nobody believes in this sort of clockwork universe any more; for one thing quantum mechanics long ago put paid to it. But even if one did, notice that the *laws* of physics as such still do not provide the constraints. It is the *cosmic initial conditions* that fix everything, and these are outside the scope of science in the clockwork universe model.

A more serious objection would be to my use of the word 'laws' (albeit within quotation marks) to describe Space Invader dynamics. After all, Space Invaders is a pretty special sort of computer game, and the word law gives the impression of something quite grand and embracing. Well, Space Invaders is but one of many applications of the principles of digital computing, and it is really these more general principles that organise the flow of electrons etc. inside the machine. Perhaps 'organising principles' is a better expression.

Some people might even object to this, arguing that it still sounds too 'fundamental'. Can one place the principles of digital electronic computing alongside the laws of physics, even though we have agreed that the latter cannot be derived from the former? Surely the laws of physics are in some sense more fundamental? I believe this is now largely a matter of taste and semantics. Is Ohm's law a law? Is Archimedes' principle a principle? Are there laws of genetics? Of social systems? Or are there only laws of super-strings? Space forbids me from reviewing the long history of attempts to argue that some so-called 'secondary' laws are every bit as fundamental as the 'primary' laws of physics. (A particularly lucid recent account has been given by Leggett (1987).) One example will suffice. In quantum mechanics, the very formalism needed to describe 'fundamental' particles demands, as Bohr showed us, the meaningfulness of the classical world of measuring devices, devices that depend on such things as Ohm's law and the second law of thermodynamics. It is then inconsistent to claim priority of the micro over the macro.

Generally, people are happy to accept that the more widespread a law or principle may be, the more fundamental it is. There is no better topical example than the burgeoning theory of deterministic chaos. (For reviews see Cvitanovic (1984), Crutchfield *et al.* (1986).) It turns out that more or less all nonlinear dynamical systems possess parameter regimes wherein the dynamics is essentially random. Although these are deterministic systems, their behaviour is so sensitive to their initial conditions that predictability is impossible. Specifically, input errors in computational models lead to output errors that diverge exponentially rapidly, so that in a finite number of e-folding times all predictive power is lost. Chaotic systems in the real world include the weather, the motion of the conical pendulum, three-body motion and certain fish and insect populations.

Although by definition it might seem that chaotic systems do not display any sytematic features, that is not the case. There is 'order in chaos'. A certain class of systems approach chaos in a sequence of precisely scaled steps, the scaling behaviour of which is described by two universal numbers - Feigenbaum's numbers (Cvitanovic 1984). Whether insect population or pendulum, the same two numbers control the escalation of complexity that is the hallmark of chaos. This universality supports the contention that Feigenbaum scaling is a fundamental law, and that Feigenbaum's numbers are fundamental constants of the physical world. These numbers cannot be derived from the traditional constants of physics, such as the speed of light and Planck's constant.

As complex systems come to be studied in more and more generality, so evidence accumulates for more of these sorts of universal systematic features. Witness the field of fractal structures. A fractal is a geometric form that is primitively and irreducibly complex, with irregularity on every scale (Mandelbrot 1982). Once a mathematical curiosity, fractals have now been identified in a thousand real-world systems, and the study of the growth of fractal structures (e.g. snowflakes) is revealing certain systematic common features. Or consider the new mathematical 'games', such as the work of Stuart Kauffman (elsewhere in this volume) and Stephen Wolfram (Wolfram 1983) on automata. Or the self-organising capabilities discovered in neural network simulations (Clark et al. 1984). These *constrained system* models of complexity are the modern frontier of theoretical model-building, imaging the real (conceded to be complex) world in a new type of mathematics, in which concepts such as non-time-reversible dynamics, algorithmic complexity and global patterning make the running. The old type of model-building dwelt on *simplicity* and relied heavily on reduction, and imaged the world (in an idealised, simplified way) by continuous curves and processes, reversible dynamics and the language of calculus.

As (quasi-universal) principles of organization begin to emerge from these theoretical studies, so they will undoubtedly find application to biological organisms, and species evolution (the latter perhaps already so – see Kauffman's article). No purpose would be served by seeking to recover these principles from the laws of physics, for they are not **in** the laws of physics. Nor would any purpose be served by calling such principles 'vitalistic', for (i) they apply to nonliving complex systems too, (ii) they are completely compatible with the laws of physics, and do not involve any new forces (only the harnessing and organization of known forces).

COMPLEXITY AS A ROUTE TO NEW PHYSICS?

So far I have been careful to stress that any newly discovered principles of organisation or complexity can stand alongside the known laws of physics, augmenting them but leaving them intact, and arguably (in some cases at least) on an equal footing as regards the designation 'fundamental'. But can we be absolutely sure that we *do* know all the laws of physics relevant to such awesomely complex systems as living organisms?

Twenty years ago it was said that there were two frontiers to modern physics – the very small and the very large. It was in the twin realms of elementary particle physics and cosmology that the truly 'fundamental' discoveries were being made. That was where genuinely *new* laws of physics would be discovered. All the stuff in between was just so much complicated detail. Nothing 'fundamentally' new would come out of it. Today we recognize that there are now three frontiers: the very small, the very large and the very complex. Armed with new mathematical techniques and computing power, physicists and their colleagues in other disciplines are tackling the very complex. Might we not expect fundamentally new physics to emerge there too? New laws and principles? One does not ask 'Does one have any right to expect Einstein's general theory of relativity to fail at some level of ultra-short distances?'. One has no right to expect it *not* to fail, and nearly all workers in this field believe it *will* fail. This continues a long and honorable scientific tradition of anticipating the limits of validity of all physical theories. Has one therefore a right to leave open the possibility that some of the known laws of physics might also fail at some level of complexity?

Many distinguished physicists have answered 'Yes' to the last question. They include many of the great names from the Golden Age of physics – Bohr, Schrödinger, Heisenberg, von Neumann, Wigner. There are also some very distinguished contemporary physicists who have answered 'Yes'. Anthony Leggett writes (Leggett 1988):

> Does the mere presence of complexity or organization or some related quantity introduce new physical laws? To put it another way, would the complete solution of the basic equation of quantum mechanics – Schrödinger's equation – for, say, the 10^{16}-odd nuclei and electrons composing a small biological organism actually give us, were it achievable, a complete description of the physical behaviour of such an organism? The conventional answer is undoubtedly yes. But what few people realize is the flimsiness – or rather, the complete absence – of positive experimental evidence for this conclusion.

In fact, physicists have a tough enough time testing Schrödinger's equation for a *three* body system, let alone a 10^{16} body system. It would clearly be unreasonable to say the least to completely *rule out* the possibility that new physics emerges at some level of complexity. We simply have no right to do this, any more than we have a right to insist that the general theory of relativity *must* be valid at all levels of size. And if it should turn out that there are irreducible, fundamentally new laws of organisation and complexity at some level, it will not only expose the attempt to explain biology in terms of individual atomic processes alone as wrong-headed (which I have concluded on the basis of the previous section's more conservative discussion), but quite simply wrong.

It is no surprise that the above list of names are all people who have worked on the foundations of quantum mechanics, for it is the attempts to interface the quantum and classical worlds consistently that has cast suspicion on the universal validity of Schrödinger's equation. Indeed, it is part of the quantum formalism that Schrödinger's equation is *not* valid when a measuring act is carried out on a quantum system. It used to be assumed that a measuring device belongs to the classical (i.e. non-quantum) world because it is large, but today we know that physical size or mass as such is irrelevant. Experiments with superconductors reveal quantum effects in systems that measure centimetres in size. Instead, the relevant parameter is complexity. A measuring device (which could be an observer's brain) must be sufficiently *complex* before a 'collapse of the wave function' (i.e. failure of Schrödinger's equation) occurs. Thus (as Bohr himself recognized (Bohr 1933)) we already know that the linear superposition of quantum atomic processes fails at some level of complexity between atom and brain.

BIOLOGY AND THE NEW PHYSICS

To date biology is rooted in the old physics, the physics of the nineteenth century. Newtonian mechanics and thermodynamics play the central role. More recent developments, such as field theory and quantum mechanics, are largely ignored. In spite of the fact that the molecular basis for life is so crucial, and that molecular processes are quantum mechanical, atoms are treated like classical blocks to be fitted together. Distinctively quantum effects, such as nonlocal correlations, coherence and phase information, let alone possible exotic departures from quantum mechanics as suggested above, are not considered relevant.

If the new physics does succeed in penetrating biology, what are the outstanding problems it must tackle? The key concepts of organisation and complexity that I have dwelt on at length are relevant both to individual organisms and to evolutionary biology as a whole. Morphogenesis, biogenesis and phylogenesis are three of the classic problems of biology to which any quasi-general organising principles find ready subject matter.

Is there a common thread running through these three problems? Indeed there is. It is called 'the arrow of time' problem. All three processes display a unidirectional growth of organizational complexity, and at first sight seem to go 'the wrong way' from the point of view of thermodynamics. (There is, in fact, no actual violation). This one-way-in-time advance in organised complexity is a puzzle to physicists because the known laws of physics are all time-reversal symmetric (with the exception of some very exotic particle physics processes that have no obvious connection with biology). Now Boltzmann showed us how the time asymmetry of the second law of thermodynamics could be reconciled with the underlying time symmetry of the dynamical laws governing the motions of individual molecules (see, for example, Davies 1974). But the second law represents a *degeneration* of order, not an advance.

A clue comes from automata and network theory. Researchers in these topics have no difficulty 'growing' structures of steadily increasing organized complexity (the entropy actually decreases), because they use *nonreversible* dynamical rules. Does this imply, as Prigogine has argued (Prigogine 1980) that we must introduce time-asymmetric modifications to the underlying laws of particle dynamics? Or should one merely turn the Boltzmann argument on its head, and use the degenerative thermodynamic irreversibility at one level to define time-asymmetric 'automata rules' that will allow *progressive* evolution at a higher level? Thus, the advance of large-scale organizational complexity could be paid for by an increase in entropy on a smaller scale, which entropy is then exported by the system (all biological systems are open) into their environment. Or is the key to be found in drawing a clear distinction between 'order' and 'organisation' here?

Space forbids me from developing these topics further, and I refer the reader to my book *The Cosmic Blueprint* (Davies 1987) for further details. But I should like to mention briefly that the exciting new field of algorithmic complexity theory (Chaitin 1988) recognises the distinction between informational complexity (and randomness) and algorithmic (roughly, computer-generated) complexity (and randomness). As our traditional ideas about entropy, thermodynamic irreversibility and the second law are all based on inormational (as opposed to algorithmic) notions of order and disorder there is scope here for a more rigourous treatment of 'organizational complexity' as opposed to mere 'order'.

One problem where randomness is invoked in a cavalier manner is in the neo-Darwinian theory of evolution. As in the thermodynamic case of random molecular shuffling, so in the biological case of random gene shuffling, there is no arrow of time. One can just as well shuffle to states of lower complexity as to states of higher complexity. An additional organising principle is needed. The principle usually invoked is natural selection. Does this then insert the all-important arrow of time that generates the unidirectional advance to states of ever-greater complexity in the biosphere? I believe the answer is not as clear as it is often made out to be.

Two niggles worry me. The first is that natural selection selects for fecundity, not complexity. Generally speaking there is an inverse relationship between the two (bacteria are far more numerous and 'successful' than people). Secondly, the unidirectional advance of biological complexity is but one example of the general tendency for the universe as a whole to advance in complexity. It seems that the cosmos started out, in the big bang, in a state of remarkable simplicity and featurelessness, and that step by step matter and energy have organized themselves into ever more complex states. This is suggestive of a much more general underlying principle of organisation that applies to both the animate and inanimate domains. Natural selection is no real help for the latter. This is not to deny that there *is* natural selection in biology; it clearly plays an important role. I am merely suggesting that natural selection alone may fail to capture the essential time-asymmetric organizing element that seems to be a quite general phenomenon. But natural selection combined with some of the ideas discussed in the foregoing may open the way to a more convincing account of the growth of biological complexity.

REFERENCES

Bennett, C.H. (1986) 'On the nature and origin of complexity in discrete, homogeneous, locally interacting systems.' *Foundations of Physics 16*, 585.

Bohr, N. (1933) *Nature* 1 April, 458.

Chaitin, G.J. (1988) *Algorithmic Information Theory*. Cambridge, Cambridge University Press.

Clark, J.W., Winston, V. and Rafelski, J. (1984) 'Self-organization in neural networks.' *Physics letts.* 102A, 207–11.

Crutchfield, J.P., Farmer, J.D., Packard, N.H. and Shaw, R. (1986) *Scientific American*, December issue, 38–49.

Cvitanovic, P. (1984) *University in Chaos*. Bristol, Adam Hilger.

Davies, P.C.W. (1974) *The Physics of Time Asymmetry*. London, Surrey University Press and Berkeley, California University Press.

——(1987) *The Cosmic Blueprint*. London, Heinemann and New York, Simon & Schuster.

——and Brown, J.R. (1988) *Superstrings: A Theory of Everything?* Cambridge, Cambridge University Press.

Farmer, D., Toffol, T. and Wolfram, S., eds. (1984) *Physica* 10D.

——eds. (1986). *Physica* 22D.

Leggett, A.J. (1987) *The Problems of Physics*. Oxford, Oxford University Press.

Mandelbrot, B.B. (1982) *The Fractal Geomtery of Nature*. San Francisco, Freeman.

Minsky, M. (1987) *The Society of Mind*. New York, Simon & Schuster.

Nicolis, G. and Prigogine, I. (1977). Self-organization in Non-Equilibrium Systems. New York, Wiley.

Prigogine, I. (1980) *From Being to Becoming: Time and Complexity in the Physical Sciences*. San Francisco, Freeman.

——and Stengers, I. (1984) *Order out of Chaos*. London, Heinemann.

Tritton, D.J. (1988) *Physical Fluid Dynamics*. Oxford, Oxford University Press.

Wolfram, S. (1983) Statistical mechanics of cellular automata. *Rev. Mod. Phys.* 55, 601–43.

9. Adaptive strategies gleaned from immune networks: viability theory and comparison with classifier systems

FRANCISCO J. VARELA, VICENTE SÁNCHEZ-LEIGHTON and ANTONIO COUTINHO

We are concerned with extracting some basic design principles that render biological immune networks adaptable to widely different environments. By adaptive we mean here that the system will *reconfigure* itself to an unspecified environment in such a way that it both maintains its ongoing dynamics and displays a behaviour that reveals a degree of inductive learning about environmental regularities. Such adaptive, learning performance is grounded on the strategy of maintaining a high rate of *replacement* of the network components.

In this broad sense, immune networks seem to have something in common with the artificial networks inspired from evolutionary principles, introduced a few years ago as Classifier Systems (CFS) by Holland (1975, 1984). CFSs were introduced in order to gain the adaptive flexibility needed to escape the 'brittleness' of other known cognitive systems, including neural networks. The idea was to provide the system with a basic biological strategy of replacement: *genetic algorithms*. One uses those network components that already seem to be providing a robust contribution to a good performance as a whole in order to generate new components by slight variations, in a manner akin to genetic recombinations, chromosomic translocations or point mutations. A number of recent studies have been concerned with the more detailed capabilities and performances of such systems to specific situations (Grefenstette, 1987; Robertson and Riolo, 1988).

In studying immune networks we have remarked that they appear to provide a different strategy for adaptiveness. Using unique somatic genetic principles for the continuous generation of diversity (novelty), immune networks introduce new components (lymphocytes) and thus update their structure by *recruitment* from a permanently existing pool of potential candidates; this recruitment is done on the basis of the global state of the network. The biological details and background for these conclusions has been presented elsewhere (Varela *et al.*, 1988).

With this in mind, our purpose here is twofold. First, our aim is to introduce a mathematical framework to express this kind of adaptive dynamical system in a precise fashion. We propose that differential inclusions and viability theory as developed by Jean-Pierre Aubin (Aubin and Cellina, 1984; Aubin, 1988a,b), might be suited to the task. In the following section we introduce this formalism and derive some consequences for the viability of immune networks. Second, in the next section, this framework allows us to compare CFS and immune approaches to viability and adaptiveness. The final remarks in the last section shed some light on the relationship between adaptation and learning, based on the unifying frame of viability theory.

Since it is not possible to cover here the background material in CFS, differential inclusions, and immune networks, we are forced to assume either that the reader is familiar with this background, or that he has access to the references which we cite in this paper.

A FRAMEWORK FOR VIABILITY THEORY

The explicit construction of viability theory was motivated by the study of the evolution of macrosystems arising typically in economics and biology. Differential equations of some relevant variables have been a very popular choice in the modelling and simulation of the dynamics of these systems. For definiteness let us assume that the system is defined by N variables $x_1(t), \ldots, x_N(t)$ constituting the state $X(t) \in \mathscr{R}^N$ at instant t. One then tries to capture the rate of variation of X by:

$$x_i(0) = a_i \tag{9.1a}$$

$$\frac{dx_i}{dt}(t) = \Phi_i(X(t), u_i(t)) \tag{9.1b}$$

We will say that Φ_i is the *dynamical law* (or rule) of the system and that u_i represents its *internal environment* (or its control, depending on context). It is assumed that Φ_i, u_i are *given*, otherwise we could not determine the precise state of the system at an instant t, which is, after all, what modelling is about.

But can we really determine the precise state of a complex system and its internal environment at t? Think of an economy, population genetics, or immune systems: do we really know how to provide the required descriptions? Will we ever know? We would say that the more complex the system is, the more negative the answers to these questions will be. In fact there is a circularity here: we tend to think of a system as complex when it forces us to answer negatively.

Can we keep something from the differential equations approach? We can agree that modelling a system implies that if we are given the state of the system and of its internal environment at t, we should be able to predict the state of the system with some success. A little bit later we should be able to predict the same for, say, $t + dt$. But once we are in $t + dt$ and we try to go on with our infinitesimal sequence of predictions, who will give us the state of the internal environment? Obviously, we don't have an equation for it, otherwise we would have included it in the system itself.

An interesting alternative to consider is to settle for less: perhaps we can specify some possible range of states, some *broad* constraints of the internal environment at $t + dt$? The aim of viability theory is to exploit this natural option. There obviously is a relation between the system and its internal environment, and this relation belongs intrinsically to the modelling of the system. We may be quite ignorant of the details of this relation, apart from being able to observe it and ascertain that the system *is* viable in its (internal) environment. Viability theory proposes to concentrate on reducing this opaqueness by expressing internal environment and permissible system states as existing within *sets*:

$$U_i(t) \in \textbf{Related\_to } (X(t)) \tag{9.2a}$$

$$X(t) \in \textbf{Allowed } \subset \mathscr{R}^N \tag{9.2a}$$

These sets are assumed as given in the modelling of the system, together with the initial state and Φ_i in equation (1). This means that, knowing the state of the system at $t + dt$, we have now a *hint* as to what state the internal environment will be in at the same instant, or at least, what it should be to permit a coexistence system-(internal) environment, *a viable trajectory for the system*.

Combining equations (9.1) and (9.2) we obtain a *differential inclusion* (Aubin and Cellina, Ch. 4 and 5, 1984):

$$\frac{dx_i}{dt}(t) \in \Phi_i(X(t), \textbf{Related\_to } (X(t))) \tag{9.3a}$$

$$\Phi_i(X(t), \textbf{Related\_to } (X(t))) := \{z|\ z\ =\ \Phi_i(X(t), u),$$

$$u \in \textbf{Related\_to } (X(t))\} \tag{9.3b}$$

In this framework the question is: for all initial states $(a_1, \ldots, a_N)$ $(= (x_1(0), \ldots, x_N(0)))$ in the **Allowed** viability domain, are there trajectories $(t, X(t))$ of the system that remain in the **Allowed** viability domain, i.e. that are **viable?** A second and related question is what subset **Viable_ Related_to** $(X(t))$ of the imposed **Related_to**$(X(t))$ is effectively used by the system to regulate its viable evolutions? We can no longer find *the* trajectory as in classical simulation, but viability is a concept at least as rich for complex systems. It represents a tight compromise between three partners in a tangled dance: a viability domain, a differential dynamics (instantaneous), and a sensible system–internal environment relation. This dance seems inevitable when considering biological systems. In the domain of design, it becomes important only if a degree of adaptiveness is required.

Casting an evolutionary network dynamics in terms of differential inclusions has two important advantages we wish to pursue. First, there is a body of results wich permits us, beyond bruteforce simulations, to understand some aspects of the behaviour of such complex systems. We will come back to this below, when we apply this framework to immune dynamics. Second, differential inclusions represent one step towards expressing an otherwise imprecise but interesting notion: learning and plasticity in 'evolutionary' systems is not due to connectivity changes (popular among neural modellers), but to changes in the *list* of participating components. In other words, classical dynamical systems do not normally include the list of participating agents as one of their variables. In a differential inclusion, although the total list can be considered still fixed, there is nevertheless an explicit way to activate or re-activate some of them (i.e. to have a non-zero concentration). This non-deterministic time-dependence of the list of network components that are (re)activated is what makes the situation not tractable by the classical tools of dynamical systems.

The viability theory approach to the interesting issue of replacement of components in a dynamical network is admittedly somewhat oblique, for we are not fully including the list as a modifiable term, but only as changing subsets of active variables. To include the list of components fully as a variable is equivalent to the following (algorithmic) procedure. The state at t is a function from the list of active participants to the real numbers:

$$\textbf{state\_at } (\text{t}): \textbf{list\_at } (t) \rightarrow \mathscr{R}$$

Let us first update the list, taking away from it those participants whose state will be zero at $t + 1$ and adding the newly recruited ones:

$$\textbf{list\_at } (t + 1) \; = \; \textbf{list\_at } (t) - \{ i \in \textbf{list\_at } (t) | \textbf{ state\_at } (t)(i)$$

$$= \; - \Phi_i(t) \} \cup \textbf{Recruitment\_at } (t)$$

then for $i \in$ **list_at** $(t + 1)$ we have the normal dynamics in case a participant was already there, and a bootstrap for the newly recruited ones:

$$\textbf{state } (t + 1) (i) \; = \; \text{if } i \in \textbf{list\_at } (t)$$

$$\text{then, } \textbf{state\_at } (t) \; = \; \Phi_i(t)$$

$$\text{else, } \textbf{Recruitment\_bootstrap\_at } (t, i)$$

This procedure captures the time dependence of the list and the state dynamics without using differential inclusions, although the connection is easily seen. However, it cannot be yet expressed in a format where we can avail ourselves of the tools of analysis. Hence for the time being we prefer to settle for the (somewhat indirect) approach offered by viability theory and a fixed list of potential participants and equations, where the active list is simulated by the participants with a non-zero state.

IMMUNE NETWORKS AS VIABLE SYSTEMS

In the case of the immune system (IS) we can make the assignments shown in Table 9.1. We concentrate here on the necessary internal environment of resting lymphocytes making possible immune dynamics. In later papers we will discuss the extension to other forms of internal environments, especially tissue markers, and eventually to external environments as incoming antigens.

Table 9.1. The internal environment of resting lymphocytes.

Viability theory	Immune networks
State $(X(t))$	Concentrations of all possible antibodies at a given instant (t)
Viability domain (**Allowed**)	Antibody concentrations bounded within some basic constraints (i.e. not being all zero, non-negative)
Internal environment	Bone marrow and available resting lymphocytes
Relation system– environment (**Related to**)	Recruitment of resting lymphocyte into network

The dynamical laws that describe immune networks have been presented and justified in detail elsewhere (Varela *et al.*, 1988). We will restrict ourselves here to a few remarks for the sake of making this paper self-contained. The variables in these equations take positive real values, but their index $i \in [1,N]$, although finite, is intended to number locations in a five- or six-dimensional manifold of molecular shapes, or shape space (Perelson and Oster, 1979). The immune system is viewed here as containing two relevant variables: the concentration of free (f) and bound (b) antibodies of a list for $i \in [1,N]$. These species interact amongst each other by their chemical affinity m_{ij} which does *not* change. At any given time the

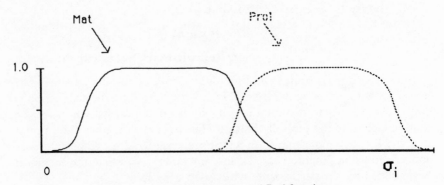

Figure 9.1. Profiles for the Mat and Prol functions.

total sum of the repertoire capable of binding at some species i represents the sensitivity σ_i of the entire system. Free antibodies are produced by the lymphocytes that advertise them on their surface by a process of maturation (**Mat**) which is a function of the sensitivity (Figure 9.1), and their activity is diminished by their binding to other species in their environment. Lymphocytes capable of producing such antibodies of species i, increase in number (if they are present) through a proliferation function (**Prol**) also on the basis of sensitivities (Figure 9.1), but they also decrease in numbers by cell death at rate k or by maturing into lymphocytes producing free antibodies. Finally, the plasticity of the system is embedded in **Recruit**: species i in the list can be added to the network (i.e. become active, have a non-zero concentration) at all times through the recruitment of lymphocytes from an available pool of resting lymphocytes constantly provided by the bone marrow, **Related_to** $(f_1(t), \ldots, f_N(t), b_1(t), \ldots, b_N(t))$. This recruitment is also made on the basis of the current sensitivity.

More precisely, for $i \in [1,N]$ the differential dynamics of immune networks is given as follows. Let

$$X = (f_1, \ldots, f_N, b_1, \ldots, b_N) \qquad \text{(state of the system)}$$

$$\sigma_{i(t)} = \sum_j m_{ij} f_j(t), \qquad \text{(sensitivity)}$$

then,

$$\frac{df_i}{dt}(t) = -\sigma_i f_i(t) + \textbf{Mat}\,(\sigma_i(t))\, b_i(t) \qquad (9.4a)$$

$$\frac{db_i}{dt}(t) = -b_i(t)[k + \textbf{Mat}\,(\sigma_i(t))] + \textbf{Prol}\,(\sigma_i(t)) b_i(t) \qquad (9.4b)$$

$$+ \textbf{Recruit}\,[i, t]$$

$$\textbf{Recruit}\,[i, t] \in \textbf{Related\_to}\,(f_1(t), \ldots, f_N(t), b_1(t), \ldots, b_N(t),) \quad (9.4c)$$

$$\textbf{Allowed} = \{X \,|\, 0 \leqslant f_i, b_i \leqslant \text{Max}; \sum_j (f_j(t)) \geqslant \text{Min}\} \subset \mathscr{R}^{2N} \quad (9.4d)$$

In this frame the viable trajectories are what could be called the *behaviour* of the immune system in its close relationships to its most relevant internal environment, the bone marrow and the pool of resting lymphocytes. Clearly, one can add another term in Equation (9.4a) for the binding to the rest of molecules from other possible environments (i.e. tissue markers, or incoming antigens), since they would correspond to species in shape space. For the time being, however, we wish to concentrate on the self-consistency of the immune system. The situation is not arbitrary, since 'antigen-free' mice do develop a normal immune network (Pereira *et al.*, 1986).

Viablity theory gives necessary and sufficient conditions for this behaviour not to be empty. It also narrows, as we already hinted, the **Related_to** () relation to a subset **Viable_Related_to**() relation, such that:

(a) **Viable_Related_to** $(X(t)) \subset$ **Related_to** $(X(t))$

(b) for almost all t during the viable trajectories we have:

Recruit $[i, t] \in$ **Viable_Related_to** $(X(t))$

This relation is built using the notion of tangent cones on **Related_to** and **Allowed** (see Aubin and Cellina, 1984). So the dynamical system, when viable, knows how to react to the constraints imposed on it (by **Allowed** by modifying the '*à priori*' **Related_to** relation to provide '*a posteriori*' **Viable_Related_to** relation.

Thus, a basic result of viability theory (see Aubin and Cellina, 1984) applied to the immune system Equations (9.4a) to (9.4d), gives the following VIABILITY THEOREM FOR IMMUNE NETWORK BEHAVIOUR:

If the sets of velocities $\Phi_i(X, U)$ *are convex and (uniformly) bounded, then the immune system has viable trajectories if and only if* **Viable_Related_to** (X) *is not empty (i.e.* **Recruit** *is defined for some i at some t). Viable trajectories are thus obtained in Equation (9.4) by using the* **Viable_Related_to** *relation instead of the* **Related_to** *relation in Equation (9.4c).*

$$\text{Recruit } [i, t] \in \text{Viable\_Related\_to } (f_1(t), \ldots, f_N(t), \qquad (9.4e)$$

$$b_1(t), \ldots, b_N(t),)$$

Convexity of the set of velocities means that if two velocities can be chosen, so can the intermediate ones. The rate of change of the system is thus chosen from a densely rich set. Metaphorically speaking, if at a given instant the system may opt between a sharp left, and a sharp right turn, this is because all the possible intermediate turning directions can also be selected. In our case, since the dynamical law is linear with respect to the recruitment process, this convexity requirement means that the set **Related_to** is also convex.

If we now interpret this mathematical result biologically, it is important to ask ourselves if the pool of available lymphocytes is convex (i.e. filling the entire shape space homogeneously). In fact, the bone marrow produces a biased repertoire of lymphocytes, according to sequence constraints in the set of germ-line genes encoding variable regions of antigens (V_H, D, J_H, V_L, and J_L) in each individual (Broder and Riblet, 1984). There are a multiplicity of V, D, and J genes from which large number of combinations result. Further, expressed antibody genes are produced by 'error-prone' recombination mechanisms which include unspecified additions and the generation of novel sequences. All of this contributes to generate an enormously large diversity in the pool of resting lymphocytes which can potentially be recruited (Tonegawa, 1983). Of some 3×10^7 new resting lymphocytes species produced every day in a mouse, 80 per cent to 90 per cent are washed off in a few days (Freitas, Rocha and Coutinho, 1986). This apparently wasteful behaviour might in fact be *explained* as a contribution to the convexity of the set **Related_to**, in order to secure viability. A final factor contributing to smoothness in this context is the typically degenerate nature of antibody interaction, which amounts to a large area of influence of each species in shape space. An interesting example of violation of this convexity condition is provided by the 'moth-eaten' mouse mutant which for genetic reasons express only a very limited spectrum of the species repertoire. Interestingly, these mice suffer from autoimmune diseases and immunodeficiency (M.A., Marcos, personal communication).

The system in its evolution projects its internal image of the environment (internal or otherwise) by building the **Viable_Related_to**(X): it says which u's are acceptable, in order to respect the viability constraint. The system thus 'knows' which are the environments that give it a chance of remaining viable, of surviving. Indeed the links between the Φ_i's, **Allowed** and **Related_to** are very strong: once the first and the second are chosen (by design), the latter will sculpt inevitably the space of the states of the environment that make viable the internal dynamics of the system. In terms of the molecular mechanisms of the immune system **Related_to**(X) and **Viable_Related_to**(X) are based on the degree of connectivity of every potential resting lymphocyte with the state of the network. The larger the **Related_to**(X) subsets, the more likely for **Viable_Related_to**(X) to be non empty, and hence the more likely for the system to be viable. Conversely, the stronger the constraints, the more likely that **Viable_Related_to**(X) will be empty and that viability will fail. All in all, it is clear how viability depends upon a right balance between the size of **Related_to**(X) and **Allowed** as expressed in **Viable_Related_to**(X) which is computed from them both. As a rule the larger the **Related_to**(X) subsets the more robust will the system be.

A perfectly deterministic differential inclusion (i.e. where **Related_to**(X) is reduced to one element) obviously verifies the convexity constraint, but is not likely to obey the viability condition (i.e. that **Viable_Related_to**$(X(t))$ is not empty) requiring in fact that velocities point inwardly relative to the viability domain. The system then 'gives no choice' to the environment for an ongoing coexistence since there are no available recruits to meet this inwardness condition.

CLASSIFIERS AND IMMUNE SYSTEMS

In the framework within which we have been working, the mechanisms of learning are sharply focused: they are expressed as the procedure **Related_to**, to engender a set wherein choices can be made for viable trajectories. This is why our computer-code-like way of writing these sets is not mere whim. The question how best to understand the kind of adaptive strategies, learning algorithms, and inference procedures that immune systems suggests to us as a viable system. One mode of answering that question is to provide an explicit representation as we have done above. Another is to compare with CFSs which are also viable adaptive systems, but which use a drastically different procedure to replace the acting components of a network system.

The similarities between immune networks and CFS are quite immediate and have already been remarked in the context of a different model of immune dynamics (Doyne Farmer *et al.*, 1987). The following Table 9.2 makes explicit the corresponding features in our framework:

Table 9.2. CFS's and Immune networks compared

CFS	*Immune networks*
message space	shape space
messages	idiotypes, antigens
classifier	antibody clone
conditions	binding, affinity
action	proliferation, maturation
strength	concentration
specificity	degeneracy
tax	cell death
? ?	network sensivity
bidding	cross-reaction
bucket brigade	? ? (network rules?)
genetic heuristic	recruitment heuristic
payoff	regulation of selfcomponents, antigens removal, etc.

Typically a classifier system may be written in roughly the same kind of non-linear equations asthe immune networks described in our approach. Following the notational equivalences just mentioned one obtains:

$$\frac{dx_i}{dt} = \alpha \sum_j m_{ij} x_j \phi\ (x_j) - \beta \sum_j m_{ji} x_j \psi(x_j) - \qquad (9.5a)$$

$$- kx_i + \textbf{Genetic}\ [i, t]$$

$$\textbf{Genetic}\ [i, t] \in \textbf{Related\_to}\ (x(t)), \qquad (9.5b)$$

or paraphrasing in the nomenclature normally used in the CFS literature:

$$\frac{dx_i}{dt} = \text{payoffs} - \text{bids} - \text{tax} + \text{genetic heuristic}$$

$$\text{genetic heuristic} = \text{Recombinations (components of the network)}$$

The (non-linear) functions Ψ, ϕ depend on the specific CFS under consideration; the reversal of the index in the summation indicates the differences between payoff and bidding. Equation (9.5a) represents the dynamical law running the system. What is not classical about such differential equations is, again, the last term, which allows the introduction of new species according to the particular heuristic chosen in Equation (9.5b), whence its differential inclusion character.

This formulation is quite different from the one proposed by Holland himself (1986). We have chosen it so as to facilitate the comparisons between both classes of viable systems. They seem to differ in two important respects. First, the basic dynamical law governing the system is obviously different in each case. This is to be expected since both laws attempt to capture two disjoint domains. This is in contradiction to the view proposed by Doyne Farmer *et al.* (1987), where both systems coincide even at this equational level. The grounds for this difference is that our formulation is not based on modelling clonal selection for immune responses as theirs is, but rather the on-going self-compatibility of an idiotype network subserving molecular identity (Varela *et al.*, 1988). In spite of this fundamentally different orientation, however, it is clear that both our formulations for the immune dynamics and a generic CFS follow the same broad family of non-linear differential equations, and can be captured as differential inclusions.

A second and more interesting distinction is at the level of the kind of **Related_to** proposed by either approach. In CFS the differential inclusion is built through genetic algorithms that essentially take neighbour components to add to the list [i] of active agents, thus updating its composition. In fact this heuristic has some limitations, and to obtain a reasonable performance, it is also necessary to have *covering* algorithms which add new components by a random search in message space, taking into account non-responsiveness to current inputs (Robertson and Riolo, 1988). The immune strategy seems to propose an interesting alternative. It will add new components by selecting from a given pool *always* provided by the bone marrow and which has some genetic bias, giving a non-uniform distribution over shape space. But selection from this pool is done on the basis of the current *global* state of the network, more specifically according to the sensitivity of the network at each point in shape space.

In its current formulations, CFSs use adaption as a learning algorithm for the performance of a specific task prescribed by the designer. In contrast, our view of the immune system is not task-oriented: we are concerned with its normal biological operation as a viable system. Thus, there is no 'informative' action of the environment on the system such as 'payoffs'; on the contrary, the internal environment 'reveals', in its interaction with the system, the mechanism the latter hosts to remain viable. There is a shift here from the normal view of learning as a playback of what was learned, to learning as adaption–discovery: the environment develops (as in photography) the image of the system.

This choice represents for us the need to move beyond the immune system as a simple defence system, to one where the internal self-consistency is the central issue. Clearly, we can extend the current model by adding provisions in the equations for interaction with an 'external' environment such as antigens. This would place the immune system and CFS, on the same performance basis, and it would be possible to define a classical learning behaviour. For instance, this model immune system could learn to 'recognize' patterns of antigen sequences presented to it. We are currently engaged in a series of tests of this capability in situations where CFS have already been evaluated.

Although we do agree that such performance tests do represent a necessary step of understanding evaluation of these learning mechanisms, it also seems important to bear in mind what living immune systems suggest: the *role of inner-environments* endowing the system with an important degree of autonomy which cannot be simply expressed as payoff in well specified tasks. Our study of immune networks suggests that even if such inner-environments can be considered as 'components' of the system, their relevance for the dynamics of the system as a whole, and the peculiar relations they maintain with it, justify a special treatment and a sharper focus.

ACKNOWLEDGEMENTS

We wish to thank Jean-Pierre Aubin and W. Danny Hillis for several insightful discussions and comments on this manuscript. We are also grateful to Thinking Machines Corporation (Cambridge, Mass.) for generously giving us the chance to run some of the immune network simulations at the Connection Machine.

Financial support for this work came from grants to Francisco Varela from the Prince Trust Fund, Shell Recherche France, and the Fondation de France (Chaire Scientifique).

REFERENCES

Aubin, J.P. (1988) *Theorie de la Viabilité*, Vol. 1. Unpublished manuscript. Paris: Université Paris Dauphine.
——(1988) Learning processes of cognitive systems: a viability approach. Unpublished manuscript. Paris: Université Paris Dauphine.
——and Cellina, A. (1984) *Differential Inclusions*. New York: Springer Verlag.
Broder, P. and Riblet R. (1984) The immunoglobulin heavy chain variable region in the mouse, 1. One hundred IgH-V genes comprise seven families of homologous genes. *Europ. J. Immunol. 14*, 922–28.
Doyne Farmer, J. Packard, A. and Perelson, A. (1987) The immune system, adaptation, and machine learning. *Pysica 22D*.
Freitas, A., Rocha, B. and Coutinho, A. (1986) Lymphocyte population kinetics in the mouse. *Immunol. Rev. 91*, 5–37.

Grefenstette, J.J. (ed.), (1987) *Genetic Algorithms and their applications* Proceedings of the Second International Conference on Genetic Algorithms. Hillsdale, J.J.: Erlbaum.

Holland, J. (1975) *Adaptation in Natural and Artificial Environments.* Ann arbor: Michigan University Press.

——(1984) Escaping brittleness: the possibilities of general-purpose learning algorithms applied to parallel rule-based systems. In: R. Michalski, J. Carbonell, and T. Mitchell (eds.) *Machine Learning: An Artificial Intelligence Approach*, vol. II. Los Altos, Ca.: Morgan Kauffman.

——(1986) A mathematical framework for studying learning in classifier systems. *Physica 22D*, 307–17.

Pereira, P., Forni, L. Larsson, E.L. Cooper, M. Heuser, C. and Coutinho, A. (1986) Autonomous activation of B and T cells in antigen-free mice. *Europ. J. Immunol 16*, 685–88.

Perelson, A. and Oster, G. (1979) Theoretical studies of clonal selection: minimal antibody repertoire size and reliability of the self/non-self discrimination. *J. theoret. Biol. 81*, 645–70.

Robertson, G. and Riolo, R. (1988) A tale of two classifier systems. Boston: Thinking Machines Corporation. Unpublished manuscript.

Tonegawa, S. (1983) Somatic generation of antibody diversity. *Nature 302*, 575–80.

Varela, F., Coutinho, A. Dupire, B. and Vaz, N. (1988) Cognitive networks: immune, neural and otherwise. In: A. Perelsson (ed.), *Theoretical Immunology*, vol. 2, pp. 391–418. SFI Series on the Science of Complexity. New Jersey, Addison Wesley.

10. The adaptation of complex systems

BERNARDO A. HUBERMAN

Brian Goodwin asked me to expound on adaptation, which he considers a significant topic for this volume, and I will therefore make some remarks that I hope will be of use to the readers. These remarks are motivated by my largely unsuccessful attempts at understanding a problem which I am sure has bothered some of you as well. I refer to the relationship between the complexity of a system and its capacity to adapt to varying constraints. While searching for a satisfactory answer, I have learned a number of things which throw some light on the issues of adaptation in open organizations and on the complexity of natural and artificial structures.

I will start by discussing adaptation and its characteristics in large embedded systems made up of many interacting agents, parts or processes. I will then introduce some notions about order, complexity and disorder, and conclude with a discussion of the original problem, which I believe remains unsolved. Noticeably absent will be any mention of adaptation in the context of biological evolution. The existence of good treatises on the subject (Mayr, 1976) and its distinct characteristics, prevent me from saying anything new about the subject. However, since there are some interesting similarities between the two subjects, other articles in this volume might further such a connection.

ADAPTATION AND THE OPEN ORGANIZATION

When analyzing the notion of adaptability, one soon discovers that there is more to it than the simple dictionary entry, which defines it as 'the ability to become suitable to a new or special use or situation'. Although capturing some of the intuitions one has about adaptive behaviour, this definition says nothing about dynamics, i.e. the time it takes for a system to respond to the presence or absence of a new constraint. And yet the latter is crucial to the successful performance of natural and artificial systems. Just as no one would consider a thermostat that takes half a year to adjust to seasonal changes in temperature to be very adaptive, neither would a social organization that responds sluggishly to relevant technological changes be considered very flexible.

There are countless examples of natural and artificial systems that can perform adaptive functions with almost no impairing delays when confronted with dynamic constraints. The eye's ability to track a moving target and the autopilot mechanism of a modern airplane provide simple examples of highly successful systems. The situation is not that clear-cut, however, in the case of organizations composed of many interacting parts or agents embedded in complex environments. When global controls are absent, a community of organisms, a firm, or a collection of interacting processes in a computational network, may not be able to respond fast enough to changing constraints or produce outputs when desired. This is because in such distributed systems the spontaneous co-ordination of activity requires sophisticated communication protocols and anticipatory decisions on the part of its agents.

The need for fast adaptive responses also appears whenever biological or computational systems have to deal with situations requiring intelligent behaviour. In the case of computers embedded in the real world, this implies an ability to operate with imperfect, delayed, and often conflicting information. It also demands real-time interactions with a physical medium in which decisions must be made before all relevant information is available. These systems, which are called *open* because their totality cannot be completely specified within a single entity or agent, are at the frontier of computer research and offer an interesting laboratory for the study of communities of interacting agents. Their main characteristics – lack of global controls and unpredictable changes in constraints – render formal theories of computation largely irrelevant for their understanding. Examples of open systems are provided by social organizations such as office systems, insect colonies, robots, and computer networks (Huberman, 1988).

A severe adaptive problem in open systems is posed by the appearance of sudden changes in the nature of the constraints acting on the network in which they are embedded. One may envision a social organization or computational ecosystem, where agents have reached a fixed point consisting of mixed strategies which are used in the solution of a complex and lengthy computational problem. This would correspond to the existence of an evolutionarily stable strategy, or ESS (Axelrod and Hamilton, 1981). The rapid availability of newer and/or faster computers or resources to the system can then introduce a different optimality criterion, for a new strategy mix may now increase the agent's performance in the solution of the problem. If the system is adaptable, one would expect it to move from the once nearly optimal mix to the new one in a very short time. This notion is illustrated in Figure 10.1, where we depict the time variation of a single constraint and the response of the system to it. Basically, the time interval,

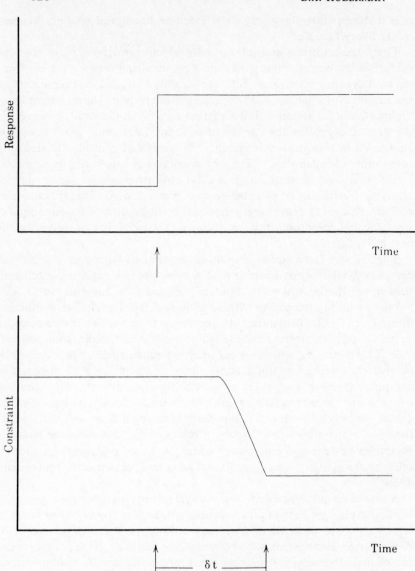

Figure 10.1. The time variation of a single constraint acting on a system and the response to it.

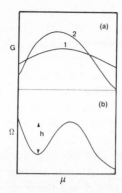

Figure 10.2. (a) Payoffs for a heterogeneous open system with two types of resources, inducing different density dependence. (b) The associated optimality function. The variable μ is the deviation from an even choice among the two strategies.

δt, between the onset of the change and the full response of the system characterizes its adaptability. For an adaptive organization, we expect this time to be very short.

In many situations of interest however, the use of better resources in a network becomes feasible only when many agents have access to them. This introduces an interesting dilemma, since small changes in the previous strategy mix can lower the performance of the system, thus preventing the appearance of spontaneous adaptive behaviour. One may then ask the following question: given such a bistable situation, with one having a higher overall payoff than the other, how does the overall system relax towards the globally optimal one?

We first note that, in the presence of such a dilemma, perfect knowledge on the part of the agents implies that the system will always stay in the relative minimum and thus not adapt. This is due to the fact that small excursions (i.e. few agents changing strategy) away from it only serve to reduce the local payoffs. Thus, knowledge about such reduction makes the agents use the same strategy mix as before. It is only in the case of imperfect knowledge that agents assume that many others have crossed over and therefore change their strategies due to their wrong evaluation of the payoffs. What is therefore needed is an estimate of the time it takes for a large number of agents with imperfect knowledge to change strategies in such a way so as to drive the system towards the new optimality minimum.

We have recently solved this problem analytically using a thermodynamic-like formulation of non-linear game theory (Ceccatto and Huberman, 1988). It amounts to the construction of an optimality function from knowledge of the density-dependent payoffs, such that its global minimum determines the long-term behaviour of the system (see Figure 10.2). In the context of computational process with no global controls and long-range interactions, we showed that under fairly general conditions, the time it takes for a system to cross over from a local fixed point which is not optimal to a global one which is, can grow exponentially with the number of agents in the system. This implies that metastable configurations in the adaptive landscape become effectively stable for large numbers of agents. When such a crossover does occur, it happens extremely fast (logarithmically in the number of agents).

Such a process is clearly illustrated in Figure 9.3, where we depict the results of a simulation (Kephart *et al.*, 1988), of a system with ten agents capable of engaging in two strategies, and with an overall payoff function of the form shown in Figure 10.2a. The choice of such a small system was dictated by the need to observe a phenomenon whose characteristic times depend exponentially in the number of processes. With all agents initially (i.e. $t = 0$) engaged in a strategy corresponding to the non-global minimum, the simulation monitored the fraction of agents, f, engaging in a given strategy mix as a function time. As can be seen, the system remained in the original configuration (i.e. $f = 0$) for a long time, and made a very sharp transition to the optimal strategy mix (which in this case corresponds to $f = 0.85$) in spite of its small size. Notice the sharpness of the transition in spite of the presence of fluctuations.

This scenario, similar to that of punctuated equilibria in evolution (Eldredge and Gould, 1972; Lande, 1985), has a number of interesting implications. A most important one is that a large collection of computational agents in an open system will not spontaneously generate adaptive behaviour when the introduction of novel constraints produces metastable configurations. In such situations, either more sophisticated procedures are needed, or a global agent has to exist in order to: (a) become aware of the advantage produced by another fixed point; and (b) to induce a co-ordinated action whereby processes simultaneously change their ESS.

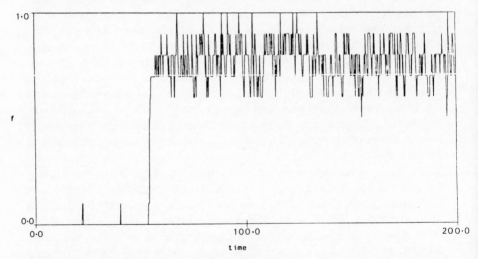

Figure 10.3. Schematic variation of physical complexity with the degree of disorder in a system.

Without either one, the evolution of open computational systems would be characterized by a number of unproductive strategy mixes with very few and rare collective transitions into more adapted ones.

When considering the implications of these results for other open organizations, an important point to notice is the effective range of the interactions among agents. For short range ones, the time to escape a metastable ESS is much shorter than the one calculated above. This is because short range interactions render the population with which an agent communicates effectively smaller, allowing the system to start nucleating the more optimal ESS in short times. For long range communications, however, the effects of local changes in the strategy mix are broadcasted to all agents in the system, thereby making the effective radius of the nucleating droplet infinite, and leading to the effects we discussed above.

ORDER, COMPLEXITY AND DISORDER

I will now turn to the problem of describing the complexity of natural and artificial systems. In contrast with the relative simplicity implied by few degrees of freedom, on the one hand, and the laws of large numbers of statistical mechanics on the other, systems whose behaviour cannot be simply understood from knowledge about the behaviour of their parts pose a number of intriguing questions about their dynamical and organizational principles. Among the important issues are the nature of organization in such systems (e.g. hierarchical versus unstructured), their dynamical properties, and the very meaning of the notion of their being complex.

Research on this topic has provided a number of powerful insights. A very appealing one, that complex systems are often hierarchical and therefore nearly decomposable, was articulated some time ago by Simon in his survey of organized structures (Simon, 1962). Hierarchies, which range from the structural of matter to the organization diagrams of social organizations, allow for an effective isolation of a given level from both the rapid fluctuations of the lower echelons and the quasi-static constraints of the higher ones. This leads in turn to dynamical processes which bear the imprint of the underlying tree structure (Huberman and Kerszberg, 1985; Teitel and Domany, 1985; Keirstead and Huberman, 1987) and which can in many cases be used for deciding on the complexity of the hierarchical system they describe (Bachas and Huberman, 1987a,b). I should also mention that the concept of hierarchy also appears in theories of biological evolution, where selection is assumed to work simultaneously and differently upon individuals at a variety of levels (Gould, 1982).

Recently, and in the spirit that allows the macroscopic states of matter to be specified by a few parameters, we introduced a complexity measure which, besides being quantifiable, encodes the relevant properties of hierarchical systems (Huberman and Hogg, 1986; Ceccatto and Huberman,

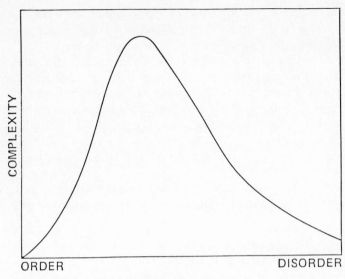

Figure 10.4.

1988). This measure, related to the diversity or lack of self-similarity of trees, stands in sharp contrast to existing information theoretic ones (Chaitin 1966), which in their present form are more suited to the analysis of messages and algorithms than to natural and artificial organizations, and cannot always be explicitly computed. Moreover, and as illustrated in Figure 10.4, this complexity measure has the property of being maximal for systems that are intermediate between ordered ones and random ones, in agreement with the intuitions one has about such structures. Within this view, the behaviour of systems which have a highly differentiated set of relationships will be harder to unravel from the properties of their parts than that of simple systems.

This notion of complexity as being intermediate between order and disorder can be illustrated with a number of simple examples. In physics one speaks of crystalline solids and gases as being relatively simple when compared with liquids or glasses. Equivalently, a magnetic system near its critical point is construed to be more complex than an ordered magnet or a paramagnet. Similarly in linguistics, the grammars of formal languages as well of those of random letters generated by the proverbial typing monkeys, possess a degree of simplicity which is in marked contrast with that of natural language.

This new measure of complexity acquires additional relevance when connected with a certain type of dynamical processes which can take place in hierarchical systems. We have rigorously shown (Bachas and Huberman, 1987a,b) that the process of diffusion in such structures is slowest for trees which lack self-similarity while fastest for both uniform and randomly multifurcating trees. Thus, in tree-like organizations, the complexity of a system determines the speed at which perturbations to a given node relax to equlibrium. This gives rise to the possibility of determining the complexity of hierarchies without resorting to their explicit organization.

In spite of this quantitative progress, there are many other notions of complexity which I have not mentioned, and which could also be considered relevant to many problems. One could, for instance, define the complexity of a system in terms of the diversity of tasks that it performs, with no reference whatever to the structure of its internal organization. In this context, a computer considered as a black box would be construed to be more complex than a bicycle, and the latter more complex than a chair. Another notion makes the complexity of a system reflect the process by which it was created (Lloyd and Pagels, 1988). Similarly, in dynamics the complexity of deterministic dynamical systems – as measured by their Lyapunov exponents – is larger when it generates chaotic trajectories than when it produces periodic motion.

In a different vein, one could also define complexity in terms of the number of distinct parts in a given structure, regardless of their organization. This is the view advocated in ecology, where a major concern is the relation between a system's complexity (measured by the number of species) and its structural stability. A common assumption is that the food webs observed in nature are the stable forms; the unstable ones having disappeared in the course of time. Studies of the stability properties of ecosystems leads to the conclusion that unstructured ones become unstable as their complexity increases (May, 1974; Pimm, 1982). Hierarchically organized ecologies, however, are the most stable against perturbations and can therefore be much larger than unstructured ones (Hogg *et al.*, 1988).

Returning to the subject of hierarchy, an unresolved question concerns the evolution of such systems and the process through which they emerge from primordially unstructured interactions. In the context of open systems, one may consider clustering different agents according to the frequency of mutal interactions, thus defining an organizational hierarchy which will presumable evolve over time-scales that are long compared to inverse interaction frequencies. This is one way in which communities such as research organizations can organize themselves in the course of time.

COMPLEXITY AND ADAPTATION

Having discussed the adaptive properties of large embedded systems and the new physical measure of complexity, I will finish by discussing the issue of their relationship. We have already seen that a desirable property of organized systems is their adaptability, which is characterized not only by the ability to satisfy variations in constraints with minimal changes in the structure, but also by response times that are short compared to those required by the performance of the system. Likewise, we have shown that hierarchical complexity, when considered as diversity, is intermediate between order and disorder. How then, do such concepts relate to one another?

The issue of their connection is still a problem with no definite answer. This is partly due to a lack of good experimental information on how the diversity of a system evolves in time as it adapts to changing constraints. There are, however, some reasonable guesses that one can make but which will have to wait for experimental verification before taking them as laws. The first one is that systems that are most *adaptable* will tend to be very complex by our measure. This follows from the fact that extreme diversity is most able to satisfy changes in constraints, a fact which has been advocated in biological evolution by a number of authors (Mayr, 1976; Jacob, 1982). Secondly, systems which are most *adapted* to given constraints will tend to have a lower complexity than *adaptable* ones, as the development of specific connections between their parts will likely lower the diversity of their structures (Huberman and Hogg, 1986). Therefore, a safe conjecture might be one which states that as a system adapts to more *static* constraints, it lowers its complexity. Notice that a corollary to this conjecture is that very adapted systems restrict their capacity for future adaptation, a phenomenon similar to that of *overspecialization* in Darwinian evolution (Gould, 1982).

This list does not exhaust all possibilities. We still have little idea about the constraints that fast responses will impose on the above conjectures, and rather than continue speculating, I am inclined to wait for some insightful data on the subject.

I end by stating my belief that careful experiments which can throw light over all these questions are within our reach. Besides the obvious organizational candidates for these tests, advances in computation should also allow for crisp simulations of systems which encapsulate the notions we have just discussed. I look forward to the time when concepts such as complexity and adaptation will be as clear and informative as those of temperature and entropy.

ACKNOWLEDGMENTS

I have benefited from many discussion and collaborations on this topic with H.A. Ceccatto, T. Hogg and J. Kephart.

REFERENCES

Axelrod, R. and Hamilton, W.D. The evolution of cooperation. *Science 211*, 1390–6.

Bachas, C.P. and Huberman, B.A. (1987) Complexity and the relaxation of hierarchical structures. *Phys. Rev. Lett. 57*, 1965–9.

——(1987) Complexity and ultradiffusion. *J. Phys. A20*, 4995–5014.

Ceccatto, H.A. and Huberman, B.A. (1988) The complexity of hierarchical systems. *Phys. Scripta 37*, 145–50.

——(1988) Persistence of non-optimal strategies, *PARC Preprint*, to appear in *Proc. Natl. Acad. Sci.*

Chaitin, G.J. (1966) On the length of programs for computing binary sequences, *J. Assoc. Comp. Mach. 134*, 547–50.

Eldredge, N. and Gould, S.J. (1972) Punctuated equilibria: an alternative to phyletic gradualism. T.J.M. Schopf (ed.) *Models in Paleobiology* San Francisco: Freeman.

Gould, S.J. (1982) Darwinism and the expansion of evolutionary theory, *Science 216*, 380–7.

Hogg, T. Huberman, B.A. and McGlade, J.M. (1988) The stability of ecosystems, *PARC Preprint*. In press, *Proc. Roy. Soc. B* (1989).

Huberman, B.A. (ed.) (1988) *The Ecology of Computation*. Amsterdam: North-Holland Publishing.

Huberman, B.A. and Hogg, T. (1986) Complexity and Adaptation. *Physica 22D*, 376–84.

Huberman, B.A. and Kerszberg, M. (1985) Ultradiffusion: the relaxation of hierarchical structures. *J. Phys. A18*, L331–L335.

Jacob, F. (1982) *The Possible and the Actual*. New York: Pantheon.

Keirstead, W.P. and Huberman, B.A. (1987) Dynamical singularities in ultradiffusion. *Phys. Rev. A36*, 5392–400.

Kephart, J.O, Hogg, T. and Huberman, B.A. (1988) Dynamics of computational ecosystems. In: *Proceedings of the Workshop on Distributed Artifical Intelligence*. Morgan Kaufmann Publishers.

Lande, R. (1985) Expected time for random genetic drift of a population between stable phenotypic states. *Proc. Natl. Acad. Sci. 82*, 7641–5.

Lloyd, S. and Pagels H. (1988) Complexity as thermodynamic depth, Rockefeller University Preprint.

May, R. (1974) *Stability and Complexity in Model Ecosystems*. Princeton, N.J. Princeton University Press.

Mayr, E. (1976) *Evolution and the Diversity of Life*. Cambridge, Mass.: Harvard University Press.

Pimm, S.L. (1982) *Food Webs*. New York: Chapman and Hall Publishers.

Simon, H. (1962) The architecture of complexity. *Proc. Am. Phil. Soc. 106*, 467–82.

Teitel S. and Domany E. (1985) Dynamical phase transitions in hierarchical structures. *Phys. Rev. Lett. 55*, 2176–9.

11. Communication and Organization in Developing Systems: a Field Viewpoint of Positional Information

L.E.H. TRAINOR

In developing systems in biology one has a complex interplay between genetic information on the one hand and the possible structures and dynamics of physical systems on the other. The view is generally held that biological systems are also physical systems, and that in the various processes taking place in development, the laws of physics and chemistry are strictly obeyed. Understanding development would then seem to involve in an intimate way an understanding of the physical organism and of the physical events taking place as part of a 'theoretical biology'. Yet this has not been the tradition in biology. As Dr L. Siminovitch has stated (in a conversational remark): 'whereas physics is theory driven, biology is not'. Is there some inherent reason why this tradition should be maintained – some intrinsic gulf which separates theoretical physics from biological understanding? We take the point of view that this gulf is at least wider than it need be and that theoretical physics provides conceptual and utilitarian frameworks which can also be of value in understanding organisms. In particular, we explore the field concept in theoretical physics and show, by examples, its unifying and explanatory value in developmental biology.

In this essay we consider in particular the question of positional information first expressed by Driesch (1901), but addressed vigorously by Lewis Wolpert (1984) and others in recent times. The question is usually stated for multicellular systems, but the problem is also there for unicellular systems as emphasized by Goodwin (1976). For multicellular systems, development takes place in a highly co-ordinated manner so that every cell seems to 'know its place', i.e. by virtue of its position in the whole organism, it functions in a certain way, whether this involves the turning on or off of appropriate genes, or whether dictated to do so by its physical and chemical circumstances, given its internal structure and form. The cell is said to possess 'positional information'. Wolpert imagines a co-ordinate system, more-or-less fixed in the body of the organism so that each cell has a location; he then poses the question, how does a cell 'know' where it is so that it can respond appropriately?

Early versions of Wolpert's theory posited that the organism managed by unspecified means to establish gradients of some chemical constituent and that genetic mechanisms were such as to respond appropriately depending on local concentrations. When a certain threshold is reached, sets of gene respond and the cell behaves or functions in an appropriate manner. This simple picture had some striking success in the early examples, but has developed some considerable problems which make it difficult to maintain. It requires not one, but many morphogenetic gradients and each of these requires cellular mechanisms with many threshold values; moreover, the identity of morphogens is problematical and the threshold mechanisms required have not been identified. It was an imaginative attempt that hasn't worked out generally enough to establish a principle.

Nevertheless, the problem remains. How, in fact, do cells coming from a common cellular ancestor organize themselves so that they not only account for overall morphology, but differentiate so that their internal states are appropriate to their positions, and thus function correctly in the organism? This is the general problem we address using a field viewpoint borrowed from physics. Since by 'field' we mean something more detailed and quantitative than the conventional use of the term in biology (e.g. the eye field or ear field in developing embryos), we first discuss the use of the term in physical contexts before discussing its value in dealing with the problem of positional information.

Arthur Koestler (1964) credits Kepler with the introduction of the field concept in relation to forces between the planets and the sun. Koestler also points out that we have no direct 'field' experience in daily life, yet have invented fields of enormous strength; if the force exerted on the earth by the sun via the gravitational field were to be supplied by a steel cable, that cable would have to be larger in diameter than the earth itself. In electro-magnetism a sophisticated attempt was made by Wheeler and Feynman (1945) to do away with the field concept and to replace it by an action-at-a-distance theory, but resort had to be made to special artefacts like perfect absorbers at infinity. It is especially the presence of radiation phenomena which seems to require the field concept as an ultimately necessary one in the fundamental physical theories of the universe. Fundamental fields of this kind (e.g. gravitational or electromagnetic), are currently regarded as entities in themselves in that they obey dynamical equations even without the presence of matter, and they manifest a field energy either in isolation or when interacting with matter.

Physicists use the term 'field' more broadly than in reference to these fundamental fields (e.g. they refer to velocity fields in hydrodynamics, to pressure fields in sound propagation, etc.). In this more general usage, a field is a space–time function obeying some dynamics and describing some property of the system. In the discussion which follows, this more general usage is employed, for example in reference to viscoelastic fields in mor-phogenesis and to developmental fields in limb transplant phenomena. But a space–time chemical concentration obeying reaction–diffusion equations also qualifies as a field. From this vantage point Wolpert's gradient (plus thresholds) model of positional information is also a field model but without detailed specification of the field dynamics. Other examples of field models (in the physicist's sense) in biology are the phase information model of Goodwin and Cohen (1969), the order–disorder model of cleavage patterns (Goodwin and Trainor, 1985) and the two models of morphogenesis discussed in the following.

With this background we now return to the question of positional information and the possible use of the field concept in answering it. In recent years it has become clear that most fields in nature are non-linear. Unfortunately, there is no well-developed, analytical approach to non-linear equations as there is for linear equations; none the less the advent of high-speed computers has made possible sophisticated calculational approaches to non-linear equations. Non-linear equations reveal a rich variety of solutions, replete with interesting mathematical behaviours such as bifurcations, symmetry-breaking, period doubling and dynamical chaos. We are in the midst of a mathematical revolution in which extremely complex behaviour can arise out of rather simple non-linear equations. The implications for biology are profound: the possibility of catching at least the essence, if not much of the substance of biological complexity, can come out of theoretical models of biological structure and function. The Waddington dream is still in its infancy but now appears realizable.

Our two examples of how field models deal with positional information also illustrate the power of non-linear dynamics in dealing with morphogenesis in biological organisms. The first example, inspired by the seminal paper of French, Bryant and Bryant (1976), is a developmental field approach to limb regeneration and transplant phenomena in amphibians (Totafurno, 1985; Totafurno and Trainor, 1987); the second, inspired by the gastrulation model of Odell *et al.* (1981), is a viscoelastic field approach to morphogenesis in the marine alga *Acetabularia mediterranea* (Goodwin and Trainor, 1985; Briere and Goodwin, 1988; Trainor and Goodwin, 1986; Hart, Trainor and Goodwin, 1988). The first example illustrates the field concept in a multicellular system, the second in a unicellular system.

In the model of Totafurno and Trainor, a developmental field is posited to account for the results of regeneration and transplant phenomena in amphibian limbs, and particularly to account for the phenomenology of supernumerary production. The question naturally arises, just what is a developmental field? In our research programme we deliberately avoided answering this question for two reasons: first, we were interested in broad rather than specific issues; second, we did not wish to get lost in detail. In part the successful application of theory in physics is to recognize that there are different levels of explanation; when studying water waves one does not reach down into the sub-nuclear grab-bag of quarks and gluons. In a similar vein, when studying general growth and form in a macroscopic structure like the limb, one does not necessarily reach down into the grab-bag of chemical reactions, substrate–receptor interactions, etc. Naturally, just as in physics, one wants to be assured that the various levels of description and explanation are eventually tied together; nevertheless, each level of explanation requires its own descriptors.

We begin our discussion with a brief statement of the problem. When the limb of a newt or salamander is severed at whatever proximo-distal level, the limb regenerates following the rule of distal transformation. After a few weeks the regenerated tissue forms a blastema which acts very much like a limb bud at that distal level. If the blastema is severed, say, from a right stump and grafted on to a left stump at the same level, the most common result is the development of the blastema into a right limb, but accompanied by two supernumeraries of left-handedness. This is referred to as a contralateral graft. In 1976, French, Bryant and Bryant (FBB) introduced an insightful model, variously referred to as the polar co-ordinate or clockface model, which accounted for these results and similar results with cockroach limbs. Essentially the FBB model is a field model, although they did not label it as such, since it assigned field values running from 1 to $12 = 0$ azimuthally about the limb both for the stump and the blastema. As healing occurs, new tissue (intercalary) fills in between blastema and stump. FBB assumed that the new cells adopted field values intermediate between stump and blastema in such a way as to minimize the morphogenetic field distance (e.g. to bridge from 1 to 4, one would go through the sequence 1, 2, 3, 4 rather than 4, 5, 6, 7, 8, 9, 10, 11, 12, 1). They called this the shortest intercalation rule. They further assumed that whenever a complete circle of values 1 to 12 occurred in the intercalary tissue, distal transformation followed (i.e. a supernumerary developed). FBB carried out their analysis by projecting the limb on to a plane perpendicular to the limb axis. Glass (1977) was quick to point out that the appearance of two left supers in the above example could be understoood topologically as an instance of the index theorem for analytic functions.

Unfortunately the polar co-ordinate model failed to account for the results obtained in ipsilateral grafts where the developing blastema was severed, turned through 180° and regrafted. For contralateral grafts, one can match two of the three morphological axes (say PD and AP but not dorsal ventral) whereas for 180° ipsilateral grafts, both the AP and DV axes are counter aligned. Moreover, in more exotic circumstances, transplants such as regeneration of mirror symmetric limbs, the model failed totally. In addition to these problems, the polar co-ordinate model had no quantitative dynamical equations and consequently had limited predictability.

Tevlin and Trainor (1985), analysed the boundary conditions in both contralateral and ipsilateral grafts assuming a field model and concluded that the radial and circumferential fields should be decoupled in the FBB model. We then undertook (Totafurno, 1985; Totafurno and Trainor, 1987) to develop a non-linear vector model of the developmental field in the intercalary region based on supers appearing as bifurcations in the field solutions. We refer to this as the TT model. In choosing a possible set of dynamical equations we followed several guidelines:

1. The field required a minimum of two components, one representing the relative amount of ventral-dorsalness, the other the relative amount of anterior-posteriorness. Our vector model still had a sufficient degree of freedom to ascribe the relative amount of distalness to the vector magnitude.

2. The field should be as smooth as possible, in keeping with the general experience with transplant phenomena. Dynamically this corresponds to including a Laplacian term in the energy function.

3. The field should be normalized in some way.

4. The field equations must be non-linear and at least of order three, so that bifurcations may occur.

5. Otherwise the equations should be as simple as possible.

This set of criteria led us to the following equations for the field dynamics in the intercalary region:

$$\frac{\partial u}{\partial t} \;=\; \nabla^2 u \,+\, \alpha u \,-\, \beta u(u^2 + v^2), \tag{11.1}$$

and

$$\frac{\partial v}{\partial t} \;=\; \nabla^2 v \,+\, \alpha v \,-\, \beta v(u^2 + v^2), \tag{11.2}$$

where u and v are respectively the morphogenetic field components along the anterior-posterior and dorsal-ventral axes. The constant α is involved in the normalization of the field and the term in which it appears acts something like a restoring force when the field is perturbed away from its equilibrium value. The constant β characterizes the strength of the non-linear term and measures the coupling between organization along the dorsal-ventral and anterior-posterior axes.

The limb in the graft region is modelled as a cylinder, and it is assumed as usual that the significant events affecting morphology are at or near the surface. The problem is then two-dimensional and the intercalary region is topologically equivalent to a flat sheet. It is, of course, important to distinguish between the geometry of the limb surface and the morphogenetic field components u and v defined with respect to that surface.

The two-dimensional intercalary sheet is bounded on the proximal side by the stump surface and on the distal side by the blastema surface tissue, as illustrated in Figure 11.1 for the contralateral graft, and in Figure 11.2 for the 180° ipsilateral graft. The length L of the intercalary sheet increases in time as healing occurs. The evidence (see Totafurno, 1985, for a detailed

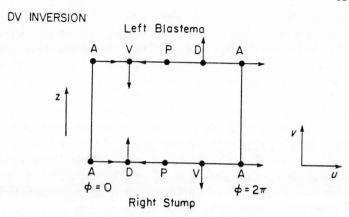

Figure 11.1. Dorsal–ventral inversion for a contralateral graft in which a left blastema has been grafted to a right stump. The intercalary region is the inside of the rectangle. The vectors show the directions of the morphogenetic field at the interfaces between the intercalary region and stump at $z = -L/2$ and between the intercalary region and blastema at $z = L/2$. Both boundaries are at the same proximo-distal level.

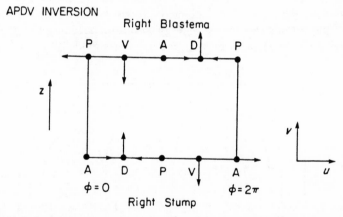

Figure 11.2. Anterior–posterior/dorsal–ventral inversion for the 180° ipsilateral graft.

discussion) supports the view that the field values at these boundaries can be taken as fixed provided well-developed blastema are transplanted. The model dynamics then corresponds to cells in the intercalary region adopting field values to bridge as smoothly as possible between the values fixed on the stump and blastema boundaries, subject to the linear and nonlinear constraints discussed above.

As mentioned above, the vector field magnitude

$$\sigma = (u^2 + v^2)^{\frac{1}{2}} \tag{11.3}$$

is a measure of 'distalness', with $\sigma = 1$ least distal and $\sigma = 0$ most distal. Just as in the FBB model, the TT vector field model does not deal directly with distal transformation *per se* but provides a sort of prepattern for distal transformation. Additional fields are required beyond the model to execute the full 3-dimensional distal program. This should not be viewed as a failure of the theory, but rather as an incompleteness since a distalizing field would be complementary and not contradictory to the field described in the model.

The results could, perhaps, be described as dramatic. Field patterns in the steady state were obtained for various values of the intercalary length L for both contralateral and ipsilateral grafts. Figure 11.3 shows the contralateral field pattern in essential agreement with FBB. The more interesting case is the ipsilateral graft for which the FBB model failed. In the TT nonlinear model, symmetric solutions are obtained for small L values, i.e. no preferred centers for distal transformation ($\sigma = 0$) occur (Figure 11.4). As L increases, the symmetry is broken by the successive appearance of two-centered, twist and four-centered solutions corresponding to two supers, no supers but a twist in the field values, and to four supers, respectively. The results are illustrated in Figures 11.5, 11.6 and 11.7. Finally, Figure 11.8. shows a plot of the free energies for the various solutions as a function of the degree of growth L. As L passes through the bifurcation value, the lowest energy solution becomes the two-centre solution. In its present form, the model is not a model of growth in itself,

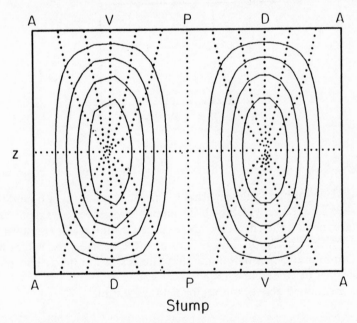

Figure 11.3. The field pattern which develops in the contralateral graft over the intercalary region in a two-dimensional representation. The solid lines are contours of constant $|\hat{\sigma}|$ with a value zero (distal tips) at the two centres. Dotted lines are lines of constant morphogenetic field direction.

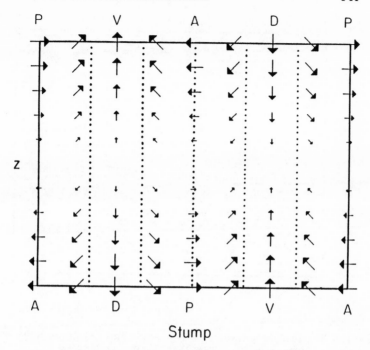

Figure 11.4. Rotationally symmetric solution for the 180° ipsilateral graft. The most distal field is on a line at $z = 0$ encircling the limb in the intercalary zone.

but rather a model for the possible organization of field values for given growth parameters.

The agreement with experiment is rather good, but difficulties remain with ipsilateral grafts (see Totafurno, 1985, for a discussion of agreements and predictions). One difficulty is that experimental results on the number of supers are scattered and one does not know to what extent artefacts such as inadequate vasculation affect the results. Complete accord might require the introduction of 'noise' into the deterministic equations of the model. Certainly there is room for further elaboration, but the model illustrates how the spirit and even substance of complicated biological phenomena can be captured in a field description.

And what of positional information? To a large extent the problem of positional information disappears or is irrelevant in a field model. The field approach is effectively a holistic approach which applies to the whole organism, or to some sub-system of an organism; local conditions are automatically in consonance with the whole – morphology is a non-local property, although there is, of course, local manifestation. This may seem like avoidance to biologists who, perhaps, tend to keep their noses rather close to the grindstone. To illustrate the validity and power of the field description, let us consider a successful analogue in the realm of conventional physics, namely the use of mean field theory in modelling physical

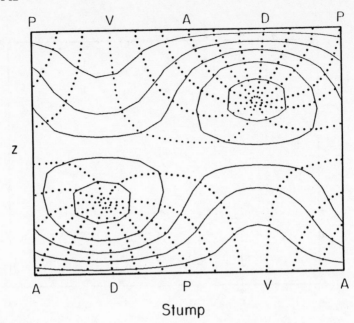

Stump

Figure 11.5. Two-centered solution for the 180° ipsilateral graft. One super is associated more closely with the stump tissue, the other with the blastema, as observed.

sytems such as crystals. At one level of description one concentrates on individual atoms, their relative locations and their electric and magnetic interactions. To bridge from the atomic detail to measureable macroscopic properties is, however, extremely difficult; but a significant and useful description can often be made by averaging out the detail in a so-called mean field description. Although such an approximation may be crude in some respects, the resulting field equations become manageable and can be used to predict such properties as order parameters, phase transitions and long-wave transport coefficients. In some sense, the 'mean field' does not exist, but is a modelling approximation to the actual electric and magnetic fields. The developmental field in our limb regeneration and transplant example is likewise a modelling approximation which transcends a complicated and detailed molecular description but captures the essence of a developing morphology.

Our second example is the use of calcium-regulated, viscoelastic fields to describe and explain the morphological changes which occur in regeneration of stem and cap in the marine alga, *Acetabularia mediterranea*. The fields in this example are more tangible than in our previous example and provide a description of morphological change based on first principles. This example is also important in that the organism is unicellular; nevertheless, the problem of positional informational of structures within the cell still exists, but is again automatically resolved in the field description.

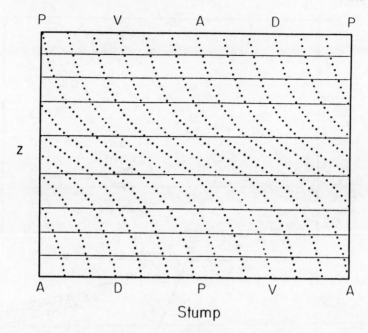

Figure 11.6. Twist solution for the 180° ipsilateral graft.

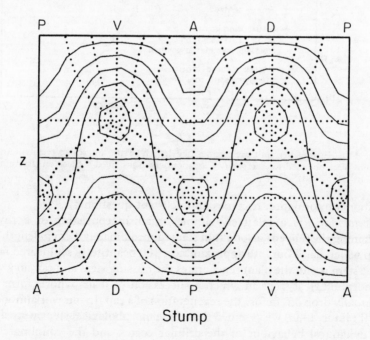

Figure 11.7. Four-centered solution for the 180° ipsilateral graft.

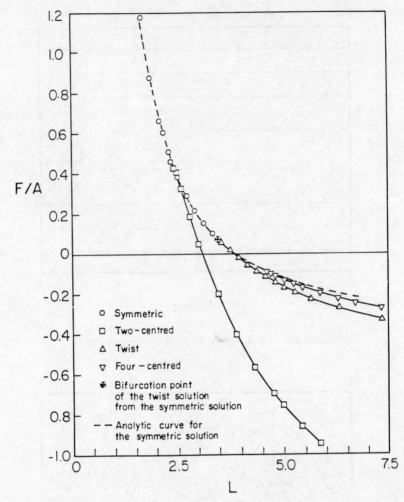

Figure 11.8. Energy per unit area versus the length of the intercalary region. Analytic solutions were obtained for the twist bifurcation which provided a check on the numerical solutions.

Positional information becomes an irrelevant question.

Acetabularia is a relatively simple organism consisting of a rhyzoid containing the nucleus, a long thin stem (approximately 1 cm in length) and a cap which flares out into approximately 100 beautifully sculptured facets. If the stem is cut, the plant regenerates the missing stem and cap, in a series of morphological steps in which a succession of 'hair' whorls form (and eventually drop off) before the regeneration of a cap. In our work (Goodwin and Trainor, 1985) we assumed that these morphological steps were due to the dynamical behaviour of the cellular cortex and the coupling of the cortex to the cell wall via the plasma lemma. Empirical evidence supports the view (Odell *et al.*, 1981) that the cortex, consisting of a network of microtubules, actinomyocin units and intermediate fibres tied together by various proteins including calcium-binding proteins, acts as a viscoelastic gel whose mechanical properties are sensitively influenced locally by the calcium ion concentration; and further, that calcium ions are released or absorbed by calcium-binding proteins depending on the state of strain in the gel. On this basis we derived a set of non-linear equations relating displacement in the gel and concentration of calcium ions. Linear stability analysis (Trainor and Goodwin, 1986; Hart, Trainor and Goodwin 1988) showed that the Goodwin–Trainor equations supported asymptotically stable solutions for structures like hair whorls and caps, with wavelength ranges dependent on such system parameters as elastic moduli, viscosities and calcium diffusion coefficients. Briere and Goodwin (1988) subsequently developed a theory of coupling to the cell wall and carried out an extensive series of calculations to show how the theory could, in principle, explain such morphological features as tip and whorl formation.

These examples show the utility of field models in describing morphogenetic change in developing organisms. In the first example, the concept of developmental fields in biology is taken seriously by introducing a field dynamic in the spirit in which physicists deal with field descriptions fo physical phenomena. The field description does not resolve problems of local detail; rather, it emphasizes the holistic nature of overall morphological development and provides quantitative results that have both explanatory and predictive power. The second example shows how, in principle, morphogenesis arises out of fundamental physical and chemical properties of biological materials and their consequent dynamics. In both examples, positional information is inherent in the field description. There is no need for interpretation of a gradient by the genes. The genes, of course, are important and have an influence on dynamics, but they do not generate or control it. Genes are not separate entities which stand apart in the developmental process, but are intimately tied into the whole machinery of development. Gene expression does not control development – it is a part of it, as evidenced in the work of T.T. Puck and his colleagues quoted in a recent note in *Scientific American* (Benditt, 1988). Puck gives clear evidence of how the cytoskeleton influences gene expression. No doubt the influence is reciprocal. A field approach bypasses the question of what controls what by treating everything as part of an integrated dynamics, but at a level of description appropriate to the phenomenon under consideration.

REFERENCES

Benditt, J. (1988) Genetic skeleton. *Scientific American* July, 40.
Briere, C. and Goodwin B.C. (1988) Geometry and dynamics of tip morphogenesis in Acetabularia. *J. Theo. Biol. 131*, 461.

Driesch, H. (1901) *Die Organsichen Regulationen*. Leipzig.

French, V., Bryant, P.J. and Bryant, S.V. (1976) Pattern formation in epimorphic fields. *Science 193*, 969.

Glass, L. (1977) Patterns of supernumerary limb regeneration. *Science 198*, 321.

Goodwin, B. (1976) *Analytical Physiology of Cells and Developing Organisms*. London: Academic Press.

Goodwin, B.C. and Cohen, M.H. (1969) A phase-shift model for the spatial and temporal organization of developing systems. *J. Theo. Biol. 25*, 49.

Goodwin, B.C. and Trainor, L.E.H. (1985) Tip and whorl morphogenesis in *Acetabularia* by calcium-regulated strain fields. *J. Theo. Biol. 117*, 79.

Hart, T., Trainor, L.E.H. and Goodwin, B.C. (1988) Diffusion effects in calcium-regulated strain fields. *J. Theo. Biol.* (in press).

Koestler, A. (1964) *The Sleepwalkers*. Markham: Penguin Press.

Odell, G., Oster, G.F., Burnside, B. and Alberch, P. (1981) The mechanical basis of morphogenesis. *Devel. Biol. 85*, 446.

Tevlin, P. and Trainor, L.E.H. (1985) A two-vector field model of limb regeneration and transplant phenomena. *J. Theo. Biol. 115*, 495.

Totafurno, J. (1985) *A Non-Linear Vector Field Model with Application to Supernumerary Production in Amphibian Limb Regeneration*. (Ph. D. thesis, University of Toronto).

Totafurno, J. and Trainor, L.E.H. (1987) A nonlinear vector field model of supernumerary limb production in salamanders. *J. Theo. Biol. 124*, 415.

Trainor, L.E.H. and Goodwin, B.C. (1986) Stability analysis on a set of calcium-regulated viscoelastic equations. *Physica 21D*, 137.

Wheeler, J.A. and Feynman, R.P. (1945) Interaction with the absorber as a mechanism of radiation. *Rev. Mod. Phys. 17*, 157.

Wolpert, L. and Stein, W.D. (1984) Positional information and pattern formation. *Pattern Formation: a Primer in Developmental Biology*. In: Malacinski and Bryant (eds.) New York: Macmillan.

12. Membrane phase transitions

FRANCISCO LARA-OCHOA

Phase transition is a property of macromolecular systems in which their elements interact. The striking characteristic of this property is that its behaviour is not merely qualitatively similar in quite different physical systems like superconductors, membrane components, ferroelectrics, etc., but often quantitatively identical. How to explain and relate the similarities in systems whose microscopic elements are as different as, for example, protein molecules and helium atoms, has been a challenge to science for many decades.

During the last year, the comprehension and description of phase transitions in regimes far from equilibrium have increased a great deal. We are now in a positon to explain many of the diverse phenomena from a unified viewpoint and even to make quantitative predictions of the behaviour of yet untested systems.

In biology, the range of phenomena and cell functions that may be explained through phase transition mechanisms is now very extensive. For instance, it was proposed (Caffrey *et al.*, 1988) that sucrose, which is almost universally present in seeds, possibly plays a fundamental role in stabilizing the membrane to prevent possible phase transitions. Goodwin (1976) has suggested that the spontaneous appearance of action potentials in *Acetabularia* may be described with a phase transition type formalism. Haken (1978) has proposed that cell differentiation might occur spontaneously in very much the same way that a ferromagnet acquires its spontaneous magnetization. Goodwin and Trainor (1980) proposed that the process for egg cleavage may be described as a series of continuous transitions, where the formalism for phase transitions may be suitable for representing the phenomenon. Shimizu and Haken (1983) proposed that the shortening of sarcomers in striated muscle and the transmission of electrical impulses in an axon are generated by a co-operative interaction between the independent elements that constitute the systems which originates a maintenance of ordered macroscopic dynamics. Yamamoto *et al.* (1984) constructed a model based on anisotropic interactions between the elements of the system to describe the circulating cytoplasmic streaming observed in *Nitella* internodal cells; Jou *et al.* (1986) have proposed a model for bacterial flagellar rotation which shows similarities to nonequilibrium phase transitions.

In this article, the phenomenology of phase transitions of phospholipids, mainly observed in membrane models and vesicles, is described. After having mentioned some of the formalisms used to describe phase transitions, some of the theoretical models of phospholipid phase transitions proposed in the literature are discussed. Moreover, some of the models concerning the excitability of membranes are presented. Finally, general conclusions on the relevance of phase transitions in development and neurophysiology are given.

In order to account for the fundamental bases of phase transitions, I refer to some of the classical models that have been used to describe the phenomenology of phase transitions in biological systems. This article does not include the concepts associated with the modern theory of critical phenomena, such as renormalization group theories. A number of excellent critical reviews on various aspects of these theories are available in the literature (Ma, 1976; Fleury, 1981; Stanley, 1971; Brush, 1967).

PHASE TRANSITION THEORIES

Physical systems as dissimilar as magnets, which can be oriented relative to the direction of an externally applied field, and phospholipids, main components of biological membranes, can be represented with different degrees of accuracy by a lattice arrangement with nearest-neighbour interactions. The simplest and most popular version of this theory is the classical model of Lenz-Ising (Brush, 1967; Hill, 1956). To describe this model, one can imagine quantized spins located at the sites of a lattice and coupled in pairs by exchange interactions. The Hamiltonian is:

$$H = \sum_{i,j,\alpha} \mathcal{J}^{\alpha}_{i,j} \, \sigma^{\alpha}_i \, \sigma^{\alpha}_j \tag{12.1}$$

where the indexes i, j specify lattice sites and the index α the components x, y, and z of the spins. The exact solution of two-dimensional Ising models has played a major part, firstly by demonstrating the existence of power laws with critical exponents different from the classical values, and secondly by showing a certain universality of these exponents with respect to various characteristics of a system like the lattice structure (see Fleury, 1981 for limitations of this universality).

Ashkin, Teller, and Potts (Toulose and Pfeuty, 1976), abandoning the idea of spins as such, proposed to assign to each lattice site m possible internal states. Thus, they postulated that there may be a non-zero interaction between two neighbouring sites if they are in different internal states, but no interaction if they are in the same state. With this concept, one can obtain new kinds of order, such as alignment, which is applicable to the study of non-isotropic fluids.

The mean-field theory relaxes the localization of each spin in the nodes of the lattice. Instead, it assumes that the behaviour of a given σ_i can be calculated from its response to the local field generated by the average spin configuration of its neighbour, i.e. by the mean field that they generate. In this view, deviations from this average, because of fluctuations, are neglected. This generalization allows the definition, at every point x of a d-dimensional space, of a field variable $\Psi(x)$. The mean field, or Ginzburg-Landau expansion of the local free-energy density, is given (Toulose and Pfeuty, 1976) by:

$$f(\Psi(X)) - f_0 = a + b\Psi + c/2\ \Psi^2 + d/3\ \Psi^3 + e/4\ \Psi^4 \qquad (12.2)$$

$$+\ ..\ +\ K \sum_{x=1}^{d} (\delta\Psi/\delta X_x)^2$$

The field variable Ψ is a quantitative measure of the development of the new phase and is called the order parameter (Landau and Lifchitz, 1967). In a ferromagnet, this parameter represents the magnetization; in a liquid crystal, it might denote the optical birefringence. The order parameter is, by definition, zero on one side of the transition and non-zero on the other side. If the order parameter increases continuously from zero in the new phase, the transition is said to be continuous or second order; if it increases discontinuously, it is said to be discontinuous or first order. The complexity of the order parameter is related to the number of components n needed to specify it completely. The coefficients a, b, c . . . of Equation (12.2) are analytic functions of T and T_c, the temperature and critical temperature respectively (see Fleury (1981) for the exact dependence of the coefficients of T and T_c). The gradient term in the expansion allows for inhomogenities in space. K is normally considered to be positive, which means that the interaction between two elements of the system is an attractive one.

The partition function can then be written as a functional integral over all spatial variations of the field variable:

$$Z = \int d\Psi \exp - \beta[\int f(\Psi)dx - H\int M(x)dx] \qquad (12.3)$$

where $\beta = 1/T$ is the reciprocal temperature, and $H\int M(x)dx$ accounts for the presence of an external field coupled to the field variable.

Landau proposes replacing the sum of all possible variations of the field variable $\Psi(x)$ with the largest single contribution, namely that which maximises the integrand, Z_{max}, i.e. minimises $\int f(\Psi(x)\ )$ dx in absence of an external field. The Gibbs free energy $G(T,H)$, for $H = 0$ is a function of the temperature T, with the form:

$$G(T) = -1/\beta \log Z_{max}\ (T). \qquad (12.4)$$

Usually, one is interested in the corresponding chemical potential, μ. Thus, taking Equation (12.2) only to the fourth order, by a simple variational procedure one has for one component:

$$\mu = b + c\Psi + d\Psi^2 + e\Psi^3 - \nabla^2\Psi$$

$$= b + c\Psi + d\Psi^2 + e\Psi^3 - \nabla^2\Psi. \tag{12.5}$$

From this minimisation can be obtained some of the so-called classical values for the critical exponents (for a wide discussion on critical parameters see Fleury, 1981).

Several other models of mean-field type exist, but in general it can be said that similar results are obtained in all the mean-field type models (e.g. they all have the same values for the critical exponents). Thus, we have the van der Waals theory of the liquid-to-gas transition, the Weiss theory of ferromagnetism, and the Bragg-Williams theory for order–disorder transitions in alloys, the Landau theory being the most succint formulation of theories of this kind.

The mean-field theory neglects correlated fluctuations in the order parameter and, of course, their considerable influence on critical behaviour. The stronger the fluctuations, the more poorly does the mean field describe the behaviour of the system as a whole. Ginzburg established a criteria for the mean-field theory to remain valid below T_c. Thus, the fluctuations $\delta\Psi$ in Ψ must remain small compared to its mean value Ψ_o.

$$(\delta\Psi)^2_{\omega\xi} \ll (\Psi^2)_{0\omega\xi}. \tag{12.6}$$

Fluctuations are correlated only over distances of order ξ, so it is proper to take the above averages over a correlation volume ω which diverges as $T \to T_c$.

PHASE TRANSITIONS OF PHOSPHOLIPIDS: EFFECTS ON CELL FUNCTION

The dynamic nature of the cell membrane arises both from a rapid biosynthetic turnover of membrane components and from the molecular movements of the components within the membrane. The membrane lipids are synthesized and degraded at a rate very much greater than the rate of generation of cells (Jones, 1979). The turnover is performed by vesicles synthesized in the interior of the cell which fuse with the plasma membrane (Lee, 1977). In this way, the membrane lipid composition may vary by responding to growth conditions or spontaneously, depending on the cell's stage of life. With this change in the compositional structure of the membrane, bacteria, yeast, or other poikilothermic organisms maintain a relatively constant fluidity, in spite of the fluctuations of temperature in their environment. It is also known that this change in composition and the consequent variation of the fluid state of the bilayer can influence the function of certain membrane proteins. For instance, lipid transition temperatures lie below or above the growth temperature in such a way that the cell may tolerate fluctuations in the external temperature. This occurs without any impairment of essential physiological processes by drastic changes in the fluidity of the plasma membrane (Jones, 1979).

In bilayers of *Tetrahymena pyriformis*, there occur changes in the lipid composition of the smooth membrane of the endoplasmic reticulum at a temperature of around 17°C (Lee, 1977). This correlates with a change in the activity of the membrane glucose-6-phosphatase. These effects are attributed to the onset of a phase transition induced by the change in the lipid composition. Another case where this takes place is during cell aggregation of *Dictyostelium discoideum*. During this stage, there is a sudden increase in the rate of interchange of phospholipids, so that the composition of the membrane changes drastically (Hase, 1982; De Silva and Siu, 1981). Simultaneously, with the change in the lipid composition, guanylate cyclase, adenylate cyclase, and other membrane-bound enzymes are activated (Bailey and Siu, 1984). It has been proposed that the change in the composition of phospholipids may be related to the onset of a phase transition (Herrera and Lara, 1988). Another case is shown through observations on the reindeer leg. In this mammal, the temperature near its feet is much lower than where the leg joins the body. To compensate for this, the phospholipid composition of the cell membranes near the feet are richer in unsaturated phospholipids than in the leg region near the body (Jones, 1979). The more double bands there are in the hydrocarbon chains, the lower the melting temperature. Thus, a relatively uniform fluidity of the cell membranes will be maintained throughout the leg, even though the temperature of the cells varies greatly over a short distance (Jones, 1979).

The lipid transition temperature in a membrane can be changed by up to 70°C (from −20°C to +50°C) by the manipulation of the fatty acid composition of the growth medium (Jones, 1979). *Mycoplasma laidlawii* grown on media enriched with unsaturated fatty acids are filamentous; if growth occurs in media supplemented with long-chain saturated fatty acids, the cells become cocoid and eventually scroll and lyse (Jones, 1979). This is possibly due to the high proportion of bilayers in the gel phase, which probably causes permeability properties to be affected and destroys the osmotic balance required for viability. The transition temperature can also be manipulated with 3-decynoyl-N-acetylcystamine, which is an inhibitor of unsaturated fatty acid synthesis (Jones, 1979). This inhibitor raises the membrane transition temperature. Very high concentrations are lethal, and cell growth ceases when the transition temperature is raised above the growth temperature. For *Bacillus stearothermophilus*, a very definite correlation exists between growth temperature, fatty acid composition, and lipid phase transition. The wild type of this organism will grow at temperatures between 37°C and 72°C, with an optimum growth temperature of 65°C (Jones, 1979). It has a very efficient mechanism for homeoviscous adaptation, through variations in the fatty-acid composition of the membrane lipids, so that the gel–lamellar phase transition always occurs approximately ten degrees below the growth temperature. A mutant, which is not able to grow above 60°C, was found to have a membrane with an approximately constant gel–lamellar phase transition at 30°C, independently of the growth temperature (Jones, 1979).

The gel to lamellar-liquid-crystal transition has been shown to be affected by local and general anaesthetics (Hill, 1974; Rowe, 1985), morphine derivatives, and anti-depressant drugs (Jones, 1979). It was generally believed that these anaesthetics work by dissolving in the lipid bilayer of excitable membranes, perturbing or expanding their structure. Membrane proteins sense the change in the lipid environment, becoming less excitable. This theory was based on the good correlation existing between the solubilities of anaesthetics in lipids and their anaesthetic potency (Miller, 1986); however, the lipid hypothesis can be questioned, because, at anaesthetic concentrations, the magnitude of the structural perturbations induced in lipids is small, probably not more than about 1 per cent (Miller, 1986). Recently, several articles have emphasized the role of anaesthetics in modifying the phase transition temperature, providing an alternative explanation on the mechanism of anaesthetics (Hill, 1974; Rowe, 1985; Jones, 1979).

The other dynamic characteristic of the membrane is the lateral diffusion in the monolayers of a bilayer of phospholipids and other membrane components. Specifically, the lateral diffusion of phospholipids is relatively fast; a phosphatidylcholine molecule in a bilayer moves at more than 50 mm/s (Jones, 1979). It has been found that a necessary requirement for the agglutination of cells by concavalin A (which is a plant haemaglutinin) is that the plasma membrane must be partially fluid, so that the conA receptors can move laterally in the membrane. For this to occur, it follows that some areas of the membrane must be above their chain-melting transition temperature. This phenomenon is essential in cell–cell interactions, and is of great importance in morphogenesis and in the maintenance of the integrity of tissues (Lara, 1984).

Phase transitions may lead to a reorganization of intrinsic components of the membrane. Phase separation of phospholipids is commonly explained by classical phase equilibria (Lee, 1979). Lateral phase separation may occur through the formation of two immiscible fluid phases, one of which is richer in one of the phospholipids (Marshall, 1978; Miller et al., 1985). It may also occur by the formation of a solid phase which separates from the fluid phase (Tokutomi et al., 1980), and through other mechanisms (for a revision see Lee, 1977).

The first step before phase separation happens is the formation of more or less lateral cluster structures made up of a lateral accumulation of the phospholipid constituents of one of the phases (Miller et al., 1985). This phenomenon, by exclusion, induces the formation of clusters of membrane-bound proteins (Marshall, 1978). The clustering of receptors in membranes is a recognized mechanism of signal reception of the cell (Bereta and Gamble, 1984).

Lateral phase separation has been found also to affect the transport properties across model membranes (Miller et al., 1985) and cell membranes (Linden et al., 1973). In the bacteria E. coli, the rate of transport of p-nitrophenyl-β-glucoside suffers a series of abrupt changes when the temperature is varied. This can be explained in terms of the energetics of transport associated with proteins in the bilayer and of the physical state of the phospholipids in the membrane (Jones, 1979).

Another consequence of lipid-phase transitions is an increased permeability to small molecules and ions. In a solid phase, the vacancies and dislocations present in an otherwise regular lattice provide favourable sites for the diffusion of small molecules. As the temperature rises, approaching the transition temperature, the number of such vacancies increases and so the permeability also increases. Lipids in the liquid–crystalline phase are less tightly packed than in the solid phase, but the disordered regions which facilitate the diffusion disappear (Lee, 1977). It should be expected that facilitated diffusion and active transport also depend on the physical state of the bilayer (Jones, 1979).

Eukaryotic membranes contain a high amount of cholesterol, as much as a molecular ratio of one, with respect to the phospholipid amount present in the membrane (Alberts et al., 1983). Cholesterol enhances the stability of the bilayer, preventing the hydrocarbon chains from crystallizing at physiological temperatures. In this way, cholesterol inhibits phase transitions and so prevents a decrease in fluidity, avoiding the subsequent physiological effects such as protein activation, changes in permeability, changes in membrane transport properties, etc.

The importance of cholesterol in maintaining the mechanical stability may be appreciated in mutants of animal cells that are unable to synthesize cholesterol. These cells lyse rapidly unless cholesterol is incorporated in the culture medium. When cholesterol is added, it is rapidly incorporated into the lipid bilayer, stabilizing it so that the cells are then able to survive (Alberts et al., 1983).

Considering that, in the presence of cholesterol, membranes exhibit the gel-to-fluid thermotropic transition at temperatures below the physiological, it becomes questionable whether such transitions can play a regulatory role in membrane functions during the normal functioning of membranes in living cells. Under these conditions, the ability of Ca^{2+} to provoke drastic shifts in the transition point of some phospholipids to a higher temperature is of particular importance (Papahadjopoulos et al., 1978).

The presence of Ca^{2+} in many physiological functions has made it be recognized as a key regulatory factor of many biological processes. At the membrane level, it has been found that Ca^{2+}, besides inducing phase transitions of phospholipids, plays a central role in membrane fusion

phenomena, such as cellular secretion and acetylcholine release in presy-
naptic nerve endings (Portis *et al.*, 1979). It has been suggested that the
complexing of Ca^{2+} with phosphatidylserine, which is an anionic phospholipid predominant in most mammalian cell membranes (Hauser and
Shipley, 1984), may provide the trigger for membrane fusion processes
(Hauser and Shipley, 1984; Wilschut *et al.*, 1985).

The kinetic results of the overall process of vesicle interaction consist of
a second-order step of vesicle aggregation, followed by the first-order
fusion reaction itself (Wilschut *et al.*, 1985). These authors emphasize that
it is not the aggregation *per se* which confers fusion susceptibility to the
vesicles. They suggest that the potential fusion capacity is due to a significant bilayer destabilization attributed to the effect of Ca^{2+} ions. Papahadjopoulos *et al.* (1977) have suggested (see also Wilschut and Hoekstra, 1984)
that the condensation of part of the outer monolayer lipids following the
binding of Ca^2 to the outside of the vesicles is crucial to this effect. This
would create phase boundaries between domains of fluid and gel-phase
lipids, constituting the focal points for fusion (Wilschut and Hoekstra,
1984).

THEORETICAL MODELS FOR PHOSPHOLIPID PHASE TRANSITIONS

In order to account for the long-range order that spontaneously arises in the
phase transitions of phospholipids, different thermodynamic models have
been proposed. Nagle (1973) has, in the formalism of the lattice approximation, included the van der Waals interactions between the chains of phospholipids as being density-dependent. For different lattice types, and
considering different strengths of the van der Waals interactions, discontinuities were obtained in the densities, related to first and second order
phase transitions. Subsequently, Nagle (1979) considered the lateral area
per chain and the surface-pressure dynamical variables, and calculated
isotherms of pressure-area. The results obtained showed that the variation
of the surface pressure or that of the area, keeping the temperature
constant, may induce phase transitions. Through modifications of the
parameter temperature, the first-order phase transitions weaken until a
second-order phase transition is obtained. Marĉelja (1974), also using the
lattice approximation to define the order parameter, took statistical
averages using the summation of all conformations of a single chain in the
field due to neighbouring molecules. He calculated the pressure–area relationship for monolayers and showed a good agreement with the experimental observations. When he varied the temperature, his pressure–area
diagrams, as those of Nagle (1975), showed a first-order phase transition.
By considering the chains of phospholipids as continuous, Jähnig (1979)
could define an order parameter, such that the model could be solved
analytically, improving the calculations of Marĉelja. Jähnig defined an

average length, l, of a chain constituted of N bonds in the direction of the surface normal, as given by the sum of the projections $\cos \theta = x_n$, such that:

$$ l = b \sum_{n=1} x_n = bx \tag{12.7} $$

where $x = 1/N \, \Sigma \, x_n$ is the average projection. He related the lateral packing density v to the average projection, considering that $\langle l \rangle / v = 1/\rho_o$ where ρ_o is the chain density. Thus,

$$ x = \frac{v}{bN\varrho_0} \tag{12.8} $$

In consequence, by using the lateral packing density of phospholipids, one may infer on the orientational order and, it follows, on the longer range order (Lara, 1988).

Subsequently, Marĉelja (1976), taking as a basis his previous model in terms of an orientational order parameter, studied the effects of non-specific interactions between proteins and lipids in membranes. Another study of this kind, also in the lattice approximation but using a Landau-Ginzburg expression for the free energy density, was proposed by Owicki *et al.* (1978). They defined the order parameter in terms of a reduced area by:

$$ u = \frac{A_f - A}{A_f - A_s} \tag{12.9} $$

where A_f and A_s represent the area per molecule in the fluid and solid phase respectively, near T_c. By including the space inhomogeneities (Owicki and McConell, 1979), the expansion of the free energy about $u = 0$ was of the form:

$$ G = T \, u^2/2 - u^3 + u^4/2 + [\nabla u]^2/2 \tag{12.10} $$

where T represents the physical temperature, scaled so that the phase transition occurs at the reduced temperature $T = 1$. They found that proteins may both change the lipid phase transition temperature and weaken the phase transition. Jähnig (1981) has criticised the use of order parameters in terms of positional order as densities or lateral areas; he proposed instead that the long-range order of protein–phospholipid interaction is better considered by using an order parameter defined in terms of the orientation of the chain. His results improved those obtained by Owicki (1978, 1979), describing the dynamical pre-transitional behaviour, the thermodynamic fluctuations, and the response functions.

While the aforementioned models describe only equilibrium phase transitions, Lara (1988) has proposed a dynamical model of non-equilibrium phase transitions. Since the most representative phase transitions in the presence of cholesterol are those induced by ion binding, the kinetics of lipid-Ca^{2+} complex formation was included in the model by appropriately defining an order parameter.

It was considered that the membrane density, v_t, was made up of lipids bounded to calcium v^{ca} and of free lipids, v^{free} (unbounded). Moreover, it was assumed that the lipids bounded to calcium are constituted by two sub-populations: lipids in the fluid phase, η^{ca}_{fluid} and lipids in the gel phase, η^{ca}_{solid}. With these considerations, the order parameter was defined as:

$$\xi = \frac{v^{ca} - \eta^{ca}_{free}}{v_t} = \frac{\eta^{ca}_{solid}}{v_t} \qquad (12.11)$$

This order parameter describes a positional order, and may be related to the orientational order by a relation of the type indicated by Equation (12.8). For values of η^{ca}_{solid} smaller than the critical point, the value of ξ equals 0. For values beyond the critical point, the maximum value reached by ξ, which represents the fraction of lipids in the gel phase, is 1.

Using the Landau approximation, the chemical potential has a form analogous to that of Equation (12.5). This chemical potential induces a mass current equal to

$$\mathcal{J} = -D \text{ grad } \mu = -D \text{ grad} \qquad (12.12)$$
$$\times \quad (b + c\xi + d\xi^2 + e\xi^3 - \nabla^2\xi)$$

where grad stands for the gradient of the chemical potential, and D represents the lateral-diffusion constant for the lipids in the membrane. The respective conservation equation was of the form:

$$\delta\xi/\delta t = -\text{ div}.\mathcal{J} + R(\xi) \qquad (12.13)$$

where the first term on the right represents the divergence of the flux and $R(\xi)$ the kinetics of the lipid–Ca^{2+} interaction.

The reported stoichiometry was taken as a basis to deduce the kinetic equation (Lara, 1988). The final equation, in non-dimensional form, which represents the dynamics of the lipids phase transition, was:

$$\delta\xi/\delta t = -\alpha\nabla^4\xi + (A + 2B)\nabla^2\xi \qquad (12.14)$$
$$+ 2B(\nabla\xi)^2 + k_f(1 - \xi)^2 - k_b\xi$$

This equation was solved by a perturbation method, using two time-scales, in two-dimensional spherical co-ordinates. The bifurcation parameter which was selected to analyse the structural stability was k_f, which is defined in terms of the specific-rate constants and the calcium concentration of the medium.

The first-order solution, which is valid only in a neighbourhood of the critical point, was of the form:

$$\xi_1(\theta, \phi) = \sum_i C_{l,m}(\tau) Y_{l,m} \qquad (12.15)$$

where the coefficients $C_{l,m}$ are dependent on the slow time τ (Lara, 1988). Thus, for different values of the bifurcation parameter, which is defined in terms of the concentrations of calcium in the medium, different spatial distributions of lipids in the liquid–crystalline phase and of lipids in the gel phase are obtained. Each spatial distribution corresponds with one of the spherical harmonics as indicated by Equation (12.5). Each stable distribution corresponds to a predominating normal mode (Huberman, 1976). Thus, domains of lipids in different phases seem to match with the aforementioned behaviour required by Papahadjopoulos *et al.* (1977) in order to give the conditions for vesicular fusion.

MEMBRANE EXCITABILITY

There are two physiologically important ways of causing excitation in a biological membrane; one is chemical and the other is electrical. Both types of excitation are characterized by a concentration–response curve which deviates from the simple Langmuir isotherm, becoming sigmoidal. The responses to electrical stimuli are not graded, but they are extremely steep; there is also a sharp threshold above which the electrical stimulus gives rise to an action potential (Blumenthal *et al.*, 1970). Chemically-excitable membranes are relatively insensitive to the membrane potential; therefore, they cannot produce an all-or-none self-amplifying excitation. Instead, their response is graded according to the intensity and duration of the external chemical signal (Alberts *et al.*, 1983).

It has been proposed that both types of excitation are mediated by a co-operative interaction between the components involved in the response. The first model proposed to describe the co-operativity of biological membranes was due to Changeaux *et al.* (1967). In the model, the biological membrane was considered as an ordered collection of repeating globular lipoprotein units or protomers organised into a two-dimensional crystalline lattice. Each protomer may have two conformational states in which one or several specific receptor sites are present per protomer. The affinity of these receptor sites toward the corresponding ligand is altered when the transition occurs from one conformational state to another. With these assumptions as a basis, Changeaux *et al.* (1976) assume that the conformational change occurs through a co-operative interaction between the protomers of the lattice. They consider the Ising approximation for the interaction energy between the protomers.

The model has been used by different authors to describe chemically or electrically excitable membranes. Most of the authors have proposed modifications in the form of considering the energy of interaction between the elements of the system. For instance, the Bragg-Williams approximation was used by Blumenthal (1970), Hill (1967), and Kijima and Kijima (1978). This approximation improves on the rigid lattice of the Ising model, used by Changeaux *et al.* (1967). The Bethe approximation has also been used

(Kijima and Kijima, 1978), as well as a modified Ising approximation (Ghosh and Sengupta, 1978), where the indirect interactions (super-exchange) between the dipoles mediated by other molecules that do not have a net electric dipole moment were considered.

More recently, a model was proposed to represent the electrical excitability of the squid axon in terms of a Ginsburg-Landau approximation (Leuchtag, 1987). This model, which is independent of that proposed by Changeaux *et al.* (1967), leads to a propagating phase transition wave accompanied by movement of ionic charge. This was interpreted as a transmembrane wave travelling along a ferroelectric unit within, and transporting ions through, the channels of the membranes.

CONCLUSIONS

The phenomena of isothermal phase transitions are specially important in eukaryotic cells. These phenomena are involved in several important cellular functions; specifically, phase transitions of lipids induced by Ca^{2+} result in the formation of spatial domains of high-melting crystalline lipid–Ca^{2+} complexes in membranes (Hauser and Shipley, 1984). This spatially heterogeneous distribution of lipids in different phases may provide the mechanism for membrane destabilization, which is the required condition for vesicle fusion. An explanation of this phenomenon has been given in terms of a non-equilibrium phase transition of the phosholipids of the membrane, induced by the Ca^{2+} ions (Lara, 1988). The importance of this phenomenon may be stressed by referring to the mechanisms of cellular excretion by exocytosis and of the release of acetylcholine in the chemical synapses.

In neurophysiology, the excitability of the membrane is undoubtedly involved in a phase transiton. However, the challenge here is identifying which are the elements which participate in the co-operative interaction. Are the co-operative elements proteins, or is the interaction a lipid–protein one which permits the co-operativity to be established? Recently, it was proposed that the excitable membrane properties of a ferroelectric type, specifically in the squid axon, are due to a crystalline ferroelectric unit which is surrounded by an intrinsic protein structure (Leuchtag, 1987).

In development, Goodwin (1976) has suggested that the spontaneous appearance of action potentials in *Acetabularia* may be the result of a phase transition. He has also proposed (Goodwin, 1983) that the calcium effects on elongation and on whorl or cap formation in this alga may be described by a state transition of the type proposed by Changeaux *et al.* (1967). He assumes (Goodwin, 1976) that, once the membrane becomes excitable, metabolic waves propagate, regulating epigenetic changes. This explanation is quite interesting, mainly in the light of more recent results observed during the fertilization of *Xenopus* eggs, where calcium waves propagating along the egg surface were detected with a photonic microscope system (Kubota *et al.*, 1987). A mechanism to explain the wave propagation has been given in terms of mechanochemical interactions (Cheer *et al.*, 1987). Calcium may thus be involved in the control of morphogenesis, not only through its modulation effect on metabolic and ionic states in cells, but also via its influence on phase transitions in membranes, and on the regulation of excitability.

ACKNOWLEDGEMENTS

I am very grateful to Esther Lara for many valuable criticisms of this work and for her help in the preparation of the manuscript.

REFERENCES

Alberts, B., Bray, D., Lewis, J., Raff, M., Roberts, K. and Watson, J. (1983) *Molecular Biolgy of the Cell*, 259–60, 351–74. New York and London: Garland Publishers.

Bailey, J.S. and Siu, C.H. (1984) Regulation of phosphatidylcholine synthesis by cAmp during early development of *Dictyostelium discoideum*. *J. Cell Biol.* **99**, 205A.

Beretta, E. and Gamble, F. (1984) A thermodynamic interpretation of clusters at the cell surface. *J. Theor. Biol.* **108**, 85–109.

Blumenthal, R., Changeaux, J.P. and Lefever, R. (1970) Membrane excitability and dissipative instabilities *J. Membrane Biol.* **2**, 351–74.

Brush, S.G. (1967) History of the Lenz-Ising model *Rev. Modern Phys.* **39**, 883–93.

Caffrey, M., Fonseca, V. and Carl Leopold, C. (1988) Lipid sugar interactions. *Plant Physiol.* **86**, 754–8.

Changeaux, J.P., Thiery, J., Tung, T. and Kittel, C. (1987) On the cooperativity of biological membranes. *Proc. Natl. Acad. Sci.* **57**, 335–7.

Cheer, A., Vincent, J.P., Nuccitelli, R. and Oster, G. (1987) Cortical activity in vertebrate eggs. I: The activation waves. *J. Theor. Biol.* **124**, 377–404.

De Silva, N.S. and Siu, C. (1981) Vesicle-mediated transfer of phospholipids to plasma membrane during cell aggregation of *Dictyostelium discoideum*. *J. Biol. Chem.* **254**, 5845–50.

Fleury, P.A. (1981). Phase transitions, critical phenomena and instabilities. *Science 211*, 125–131.

Ghosh, P.K. and Sengupta, D., (1978) A model for the temperature dependence of membrane excitability. *J. Theor. Biol.* **127**, 341–59.

Goodwin, B.C. (1976) *Analytical Physiology of Cells and Developing Organisms,*pp. 203–8. London, New york, San Francisco: Academic Press.

Goodwin, B.C., Skelton, J.L. and S.M. Kirk Bell (1983) Control of regeneration and morphogenesis by divalent cations in *Acetabularia mediterranea*. *Planta* **157**, 1–7.

Goodwin, B.C. and Trainor, L.E.H. (1980) A field description of the cleavage process in embryogenesis. *J. Theor. Biol.* **86**, 757–770.

Haken, H. (1978) *Synergetycs* (second enlarged ed.) pp. 313. Berlin, Heidelberg, New York: Springer-Verlag.

Hase, A. (1982) Changes in phospholipid compositions during the development of *Dictyostelium discoideum. Arch. Biochim, Biophys. 219*, 21–9.

Hauser, H. and Shipley, G.G. (1984) Interactions of divalent cations with phosphatidylserine bilayer membranes. *Biochem. 23*, 34–41.

Herrera, A. and Lara, F. (1988) A cooperative model for an enzyme membrane bound. In preparation.

Hill, M.W. (1974) The effect of anaesthetic-like molecules on the phase transition in smectic mesophases of dipalmitoyllecithin, I. The normal alcohol up to C = 9. *Biochim. Biophys. Acta. 356*, 117–24.

Hill, T.L. (1956) *Statistical Mechanics*. New York: McGraw Hill.

——(1967) Electrical fields and the cooperativity of biological membranes. *Proc. Natl. Acad. Sci. 58*, 111–114.

Jähnig, F. (1979) Molecular theory of lipid membrane order. *J. Chem. Phys. 70*, 3279–90.

——(1981) Critical effects from lipid–protein interaction in membranes. I Theoretical description. *Biophys. J. 36*, 329–45.

Jones, M.N. (1979) The thermal behaviour of lipids and biological membranes. In: M.N. Jones (ed.). *Biochemical Thermodynamics*, Amsterdam, Oxford, New York: Elsevier Scientific Publishers. pp. 185–223.

Jou, D., Perez-Garcia, C. and Llebot, J.E. (1986) Bacterial flagellar rotation as a nonequilibrium phase transition. *J. Theor. Biol. 122*, 453–8.

Kijima, H. and Kijima, S. (1978) Cooperative response of chemically excitable membrane. *J. Theor. Biol. 71*, 567–585.

Kubota, H.Y., Yoshimoto, Y., Yoneda, M. and Hiramoto, Y. (1987) Free calcium waves upon activiation in *Xenopus* eggs. *Devl. Biol. 119*, 129–36.

Landau, L. and Lifchitz, E. (1967) *Physique Statistique*. pp. 512–47. Moscow: Editions MIR.

Lara-Ochoa, F. (1984) A generalized reaction diffusion model for spatial structure formed by motile cells. *Biosystems 17*, 35–50.

——(1988) Dynamical model of phase transiton of phospholipids induced by Ca^{2+} ions. Sent to publication.

Lee, A.G. (1977) Lipid phase transitions. *Biochim. Biophys Acta. 472*, 237–81.

——(1977) Lipid phase transitions and phase diagrams, II. Mixtures involving lipids. *Biochim. Biophys. Acta. 472*, 285–344.

Leuchtag, H.R. (1987) Phase transitons and ion currents in a model ferroelectric channel unit. *J. Theor. Biol. 127*, 341–359.

Linden, C.D., Wright, K.L., McConell, H.M. and Fox, C.F. (1973) *Proc. Natl. Acad. Sci. 70*, 2271–75.

Ma, S. (1976) *Modern Theory of Critical Phenomena*. London, Amsterdam; Mills; Ontario, Sydney, Tokyo. Benjamin.

Marĉelja, S. (1974) Chain ordering in liquid-crystals, II. Sturcture of bilayer membranes. *Biochim. Biophys. Acta 367*, 165–76.

——(1976) Lipid-Mediated protein interaction in membranes. *Biochim. Biophys. Acta 455*, 1–7.

Marshall, A.G. (1978) *Biophysical Chemistry* New York, Santa Barbara, Chichester, Brisbane, Toronto; J. Wiley, pp. 25–32.

Miller, K.W. (1986) Are lipids or proteins the target of general anasthetic action? *Trends in Neurol. Sci.*, February, 49–51.

Miller, A., Schmidt, G., Eibl, H. and Knoll, W. (1985) Ca^{2+}-induced phase separation in black lipid membranes and its effect on the transport of a hydrophobic ion. *Biochim. Biophys. Acta 813*, 221–9.

Nagle, J.F. (1973) Theory of biomembrane phase transitons. *J. Chem. Phys. 58*, 252–64.

——(1975) Chain model theory of lipid monolayer transitons. *J. Chem. Phys. 63*, 1255–61.

Owicki, J., Springatte, M.W. and McConell, H.M. (1978) Theoretical study of protein–lipid interactions in bilayer membranes. *Proc. Natl. Acad. Sci. 75*, 1616–19.

Owicki, J.C. and McConell, H. (1979) Theory of protein–lipid and protein–protein interactions in bilayer membranes. *Proc. Natl. Acad. Sci. 76*, 4750–4.

Papahadjopoulos, D., Vail, W.J., Newton, C., Nir, S., Jacobson, K., Poste, G. and Lazo, R. (1977) Studies on membrane fusion III. The role of calcium-induced phase changes. *Biochim. Biophys. Acta. 465*, 579–98.

Papahadjopoulos, D., Portis, A. and Pangborn, W. (1978) Calcium induced lipid phase transitions and membrane fusion. *Ann. N.Y. Acad. Sci. 308*, 50–66.

Portis, A., Newton, C., Pangborn, W. and Papahadjopoulos, D. (1979). Studies on the mechanism of membrane fusion: evidence for an intermembrane Ca^{2+}-phospholipid complex. Synergism with Mg^{2+} and inhibition by spectrin. *Biochem. 18*, 780–90.

Rowe, E.S. (1985) Thermodynamic reversibility of phase transitons. Specific effects of alcohols on phosphatidylcholines, *Biochim. Biophys. Acta. 813*, 321–30.

Shimizu, H. and Haken, H. (1983) Cooperative dynamics in organelles. *J. Theor. Biol. 104*, 261–273.

Stanley, H.E. (1971) *Introduction to Phase Transitions and Critical Phenomena.* New York: Oxford University Press.

Tokutomi, S., Ohki, K. and Ohnishi, S. (1980) Proton-induced phase separation in phosphatidylserine/phosphatidylcholine membranes. *Biochim. Biophys. Acta. 596*, 192–200.

Toulose, G. and Pfeuty, P. (1978) *Introduction to Critical Phenomena and the Renormalization Groups* pp. 18–50. New York, Chichester, Brisbane, Toronto: J. Wiley.

Wilschut, J. and Hoekstra, D. (1984) Membrane fusion: from liposomes to biological membranes. *Trends in Biochem. Sci. 9*, 479–483.

Wilschut, J., Nir, S., Scholma, J. and Hoekstra, D., (1985) Kinetics of Ca^{2+}-induced fusion of cardiolipin–phosphatidylcholine vesicles: correlation between vesicle aggregation bilayer destabilization and fusion. *Biochem. 24*, 4630–6.

Wilschut, J., Scholma, J., Bental, M., Hoekstra, D. and Nir, S. (1985) Ca^{2+}-induced fusion of phosphatidylserine vesicles: mass action kinetic analysis of membrane lipid mixing and aqueous content mixing. *Biochim. Biophys. Acta. 821*, 45–55.

Yamamoto, Y, Mioh, H. and Shimizu, H. (1984) A kinetic Ising model for dynamical behaviouir of anisotropic system – its significance in streaming phenomena. *J. Theor. Biol. 109*, 373–91.

13. Coherent excitations and the physical foundations of life

The physical world is dominated by the second law of thermodynamics (i.e. the degradation of useful energy ultimately into heat, or random molecular motion). At the same time, order dissolves into disorder, a measure of which is entropy. As undergraduates, we were all taught that for any real chemical reaction, entropy always increases, and there is no possibility of reversing the change.

The biological world, by contrast, is capable of maintaining and re-producing organization on a macroscopic scale from a flow of energy and matter. The fundamental problem of life is that of how it can transform energy so efficiently, and, at the same time, organize matter so fruitfully. In this paper, I will try to show that energy transformation and living organization are mutually dependent. In other words, the dynamic organiz-ation of living matter gives rise to efficient energy transfer, and in turn the processes of energy transformation are instrumental in organizing living matter.

This is a purely speculative essay, written mostly for my own benefit as a biologist's attempt to understand the theory of coherent excitation (Fröhlich, 1980) and to see how it may be involved in biological processes that, so far, have resisted conventional explanations. My interest in this subject stems first of all from a belief that this new way of thinking about living organization and function plays a pivotal role in the global phase transition that is already taking place across many scientific and non-scien-tific disciplines: the emphasis on integration over fragmentation, on co-operation rather than competition, on dynamics and process in place of the static and mechanical, on non-linear distributed interrelationships and emergent properties of collective wholes instead of linear, unidirectional or hierarchical control of incidental parts. Most important of all, in acknow-ledging a reality in which we as scientists and human beings participate, contemporary science re-establishes our kinship with nature; thus ending centuries of abstractions which alienated sicence from humanity and humanity from nature (see Ho, 1988a).

There is of course, a second reason why we have to look at living

organization in a new light: the analytical tradition has consistently failed
to address the issue of how organized wholes can arise. Let me begin by
referring to some actively researched areas in biology where a great deal of
conventional biochemical and cell-biological knowledge has already ac-
cumulated about the parts, and yet somehow, the soul of understanding still
eludes us.

The first is the biology of vision. Photons are absorbed by molecules of
the visual pigment, rhodopsin, situated in the special membrane stacks in
the outer segment of a special rod-shaped cell in the retina. This gives rise
to a nerve impulse at the opposite end of the cell. It is estimated that the
human eye is sensitive to a single quantum of light, which is absorbed by
a single molecule of rhodopsin. The energy of the ensuing impulse is about
a million times that of the photon absorbed. Where does the extra energy
come from? Biochemists appeal to the molecular cascade, which does
indeed exist. It is a sequence of enzymic reactions in which a receptor
protein, in this case the visual pigment rhodopsin, activates a second
protein, transducin, which in turn converts another protein from inactive
into active form. Because the final protein in this cascade is an enzyme,
phosphodiesterase (which can split many molecules of its substrate, cyclic
guanosine monophosphate, or cGMP), it is easy, in principle, to see how an
initial effect can be greatly multiplied. The cGMP keeps sodium channels
open, but the split, non-cyclic GMP is ineffective; with the result that the
sodium channels close, giving rise to hyperpolarization of the cell
membrane which is supposed to initiate the nerve impulse (see Stryer,
1987). There are notable caveats in this account. For one thing, the component steps have time constants which are too large to account for the
rapidity of the neuronal response, estimated to be of the order of 10^{-2} s.
Thus, it takes approximately 10^{-2} s to activate *one* molecule of phosphodiesterase after photon absorption, 10^{-3} s for the phosphodiesterase to
hydrolyze one molecular of cGMP, and another 10^{-2} s for the split cGMP
to diffuse from the outer segment to the other end of the elongated rod cell.
Another difficulty is that the single photon-activated rhodopsin molecule
activates about 500 molecules of transducin non-enzymatically. There is no
mechanism within conventional biochemistry which explains that.

The *instantaneous transduction* and *amplification* of input signals are
characteristic of living organisms, which implies that they have a way of
storing energy in a form of readiness to respond very quickly to incoming
stimuli (and also explains why they can be very sensitive to specific, low-
intensity cues).

Another area is the biology of muscle contraction. The molecular components involved are known in detail, as is the mechanism of contraction
due to the alternating myosin and actin fibres sliding past each other by
molecular treadmilling between myosin head groups and serial binding
sites on the actin fibre. Energy comes from the hydrolysis of the high-

energy intermediate, ATP, into ADP and inorganic phosphate, Pi. This energy is used simultaneously to make and break innumerable inter-molecular bonds between myosin head groups and the actin fibres in a co-ordinated way within split seconds. To be precise, in order to lift only one gram weight through one centimetre requires the energy from 10^{14} molecules of ATP. How can the energy originally stored in 10^{14} molecules become released and channelled efficiently into muscle contraction before it dissipates into heat? This problem is multiplied at least a thousand-fold when we consider the co-ordination required in lifting one's arm. Thus, another characteristic of living organism that is in need of explanation is that of *long range coherence in energy transduction* (i.e. the rapid co-ordination of energy transformation over macroscopic distances).

The third area is that of *cellular differentiation*: how, say, a cell with specialized functions (such as that in the lining of the small intestine) can acquire its particular structure and morphology; how, for instance, dif-ferent proteins can know their address, whether to go to the apical membrane or to the basolateral membranes. A lot has been written about specific carrier particles and proteins, but this only pushes the problem into infinite regress because how do the carriers know *their* address?

Finally, the greatest enigma of all must be how *an egg or seed can develop into an organism*; how, for example, a fruit-fly embryo comes by its anterior and posterior ends, and its particular configuration of repeated body parts. Much has been done on the genetics of pattern formation, and cloned genes have been used to locate specific transcripts in the early embryo. But the problem of how spatial organization arises cannot be reduced to the mere turning on of genes which control other genes. Here, we get into the same difficulty of infinite regress as in the case of cellular differentiation.

I shall now go on to describe some relevant theoretical background to coherent excitation, and then suggest how it may contribute to solving the problems of living organization that I have mentioned.

RESONANCE AND THERMODYNAMICS OF THE LIVING STATE

Coherent excitation is intimately tied up with the concept of *resonance*. Resonance refers to something concrete and well-known in chemistry, although it has not yet gained the wide attention it deserves. Chemical bonds, when excited, vibrate at characteristic frequencies; and two or more bonds which have the same intrinsic frequency of vibration will resonate with one another. More importantly, the energy of vibration can be trans-ferred through relatively large distances from one molecule to the others. Resonant-energy transfer occurs very rapidly, typically in 10^{-14} s, whereas the vibrations themselves die down, or thermalize, in 10^{-9} s to 10^{1} s. Resonant energy transfer is 100 per cent efficient, and resonating molecules, like people, can attract one another. The importance of this mechanism cannot be overemphasized. In conventional chemistry, energy transfer occurs only at short distances (about 10^{-10} m) by collision, it is inefficient because a lot of the energy is dissipated as heat, and specificity is low, as non-reactive species could collide with each other as often as reactive species. By contrast, resonant energy transfer occurs not only at considerably larger distances (Arnott and McDowell, 1958) and without loss; it is also very specific, being determined by the frequency of the vibration itself.

One of the first people to draw attention to this method of energy transfer between molecules in biology was McClare (1971), who reformulated the second law of thermodynamics for reactions involving single molecules in place of the traditional approach, which applies to ensembles of molecules.

It is commonly believed that the second law is a statistical law that applies only to systems consisting of large numbers of particles. But a biological cell typically contains one or two molecules of DNA, and only one to a few molecules of important regulatory proteins (such as the *lac* operon repressor). It is nonetheless capable of responding with utmost efficiency to the appropriate stimuli. In the biological domain, therefore, the second law has to be reformulated to allow it to be applied to reactions involving single molecules.

This can be done by introducing the notion of a characteristic time-scale. Consider a system at equilibrium at temperature θ within an interval of time τ. The energies contained in this system can be partitioned into *stored* energies versus *thermal* energies. Thermal energies are those that exchange with each other and reach equilibrium in a time less that τ, so thay give the typical Boltzmann distribution characterized by the temperature θ. Stored energies are those that remain in a different distribution for a time greater than τ, either in a distribution characterized by a higher temperature, or such that higher energy states are more populated than states of lower energy. So, *stored energy is any form which does not thermalize in the interval* τ.

The second law can then be restated as follows: it is impossible to convert thermalized energy into stored energy, and thermalized energy is unavailable for work. This has the consequence that if energy is stored in single molecules (as it is), then those biological processes in which useful work is done cannot be using a mechanism involving thermalization of stored energy. Furthermore, since thermal exchange is the common characteristic of all chemical machines, then at least *some* biological processes do not involve conventional chemical machines.

Another issue arising from the second law is that of the supposed irreversibility of all real processes, and the conclusion that a process can only be reversible when it is infinitely slow. A reversible process obviously operates at maximum efficiency when little or no useful energy is lost. As is well-known, when a gas expands rapidly, say, against a piston in the automobile, it does less useful work than if it is constrained to expand slowly. This is basically because the energies of the molecules can only equilibrate at a finite rate.

With McClare's formulation, however, a reversible thermodynamic process merely needs to be slow enough for all thermally-exchanging energies to reach equilibrium (i.e. slower than τ) which can in reality be a very short period of time. So, high efficiencies of energy conversion can be attained in thermodynamic processes which occur quite rapidly. On the other hand, there can also be a process carried out so quickly that it is reversible. In other words, provided the exchanging energies are not thermal energies in the first place, but remain stored, then the process is limited only by the speed of light. Irreversibility is not inevitable, and hence *entropy production is not necessary to all processes.* This is, of course, implicit in the classical formulation, $\delta S \geqslant 0$, for which the limiting case is $\delta S = 0$.

There are thus two ways of doing useful work: slowly with respect to τ (which need not be very slow), or quickly, before thermal exchange can take place. In other words, if stored energy could be released in some specific form and then converted into another stored form so quickly that the probability of it exchanging with thermal energies is negligible, then useful work can be done with it. In resonant energy exchange, since it occurs at least five order of magnitude faster than vibrational decay, the nuclei involved in the exchange are virtually at a standstill, and useful work *can* be done.

I have discussed this at some length in order to stress the possibility that biological processes do not involve statistical ensembles of molecules as in conventional chemical mechanisms; instead, *all molecules of the system can act coherently, without special coercion or control.*

McClare (1972) used the principle of resonant energy transfer to explain the efficiency of muscle contraction, where it has been shown that the energy released in the hydrolysis of ATP is almost completely converted into mechanical energy. In other words, there is no increase in entropy in the reaction. McClare suggested that the energy in the hydrolyzed ATP is transferred resonantly to an oscillator in the myosin head group, which in turn attracts a similar oscillator in the actin fibre. As they approach each other, the head group tilts; work is done to make the myosin and actin filaments slide an infinitesimal distance over each other. This de-excites the coupled oscillators and the two proteins are released ready for a new cycle (see Figure 13.1). Not so long ago, Hibbard *et al.* (1985) induced rapid photolysis of ATP in single muscle fibres with laser-pulse to initiate cyclic cross-bridge formation, so as to monitor the accompanying changes in stiffness and tension in the fibre and relate them to the elementary mechano-chemical events of the energy transducing mechanism. Their results suggest that the formation of the myosin–actin complex is coupled

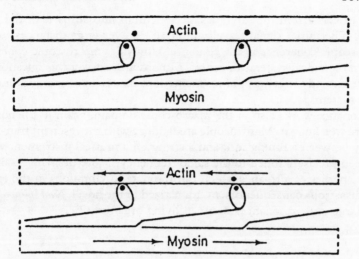

Figure 13.1. McClare's (1972) model of muscle contraction based on coupled oscillators. The oscillators are represented as black dots on the myosin head groups and the actin fibres. Arrows indicate the direction of sliding movement.

to the release of inorganic phosphate from ATP in a reaction that is readily reversible; in other words, a reaction that generates no entropy, much as McClare had predicted.

(It is of interest that in photosynthesis, where light energy is rapidly and efficiently transformed into chemical energy, recent work suggests that after photon absorption, the first step of the charge separation process is a readily reversible reaction that takes place in 10^{-14} s (Fleming *et al.*, 1988). It seems highly likely that resonant energy transfer is involved here too.)

As to the question of how the different myosin–actin complexes make and break in a co-ordinated way along the length of the fibres, McClare's answer was that the energy from excited ATP molecules can activate and attract other ATP molecules by resonant energy travelling along the length of the actin fibres themselves. The actin fibres are in fact composed of long molecules of tropomyosin, coated with actin dimers and another protein, troponin. As a result of stimulation by a nerve, a high concentration of Ca^{2+} is released locally, which binds to troponin, making it relax from the tropomyosin, and enabling the energy from ATP hydrolysis to be conducted as phonons along the tropomyosin cable in order to activate other actin–myosin oscillators as well as ATP. In this way, the spatial co-ordination of function may be achieved.

In fact, the release of Ca^{2+} is also co-ordinated over the whole muscle cell, which is a very large syncytium, of the order of 10^{-3} m in length. The initial firing of the nerve ending at the neuromuscular junction triggers an action potential to spread into a complicated series of membranous folds that extent from the muscle cell membrane to surround each small bundle of actin and myosin fibres. This in turn leads to the simultaneous, massive efflux of Ca^{2+} from the sarcoplasmic reticulum into all parts of the cyto-plasm. The initial signal travel within 10^{-3} s over large distances to give macroscopic coherence in the response. How does macroscopic coherence arise in biological systems in general? In order to answer this question, we need to extend the concept of resonance into the realm of energy storage in living systems.

Resonance works also in the macroscopic domain, again in phenomena that are well known. Many people are familiar with the resonant transfer of energy between a tuning fork and a string on a musical instrument which has been precisely tuned to the same frequency. More dramatic examples are the bodies of airplanes which can resonate to vibrations in the engine with disastrous consequences, as dramatized in the novel, *No Highway*. Do such large-scale resonances occur in living organisms?

RESONANCE AND LONG-RANGE COHERENCE

Biological systems are made up predominantly of dipolar molecules (i.e., molecules with well-separated positive and negative charges. Proteins, RNA and DNA are giant dipole molecules. In addition, the cell membrane is a macroscopic dipolar structure with its inner side at about -100 mV relative to the outside. There is thus an enormous potential gradient across the membrane of some 10^7 V/m. Also characteristic of biological materials is their elasticity, which means they undergo elastic deformation when polarized. The feedback between electrical polarization and elastic de-formation gives rise to highly non-linear vibrations. Because the dipoles are organized as cells and tissues, they also readily interact with one another, with the result that the properties which emerge from the whole are quite different from those of the isolated parts. Thus, extracted molecules in solution may not tell us much about their role within the organized cell, and this has far reaching implications for empirical investigations into living organization in general (see below).

Fröhlich (1968; 1980) put forward a theory which suggests that if energy is supplied locally to one component in a biological system, it is soon shared out by resonance-like transfer to all the components which have similar (though not necessarily identical) frequencies of oscillations. In other words, the energy is not completely thermalized, or lost, as heat; but is instead shared quickly to a collective. If the energy supply exceeds a critical level, then all the components will concentrate the energy of vibration at the lowest collective frequency (a process analogous to the Bose-Einstein con-densation that occurs at low temperatures in physical systems). That is, they will undergo coherent oscillations which may extend spatially over large regions. The easiest way to envisage this is with respect to the cell membrane. During the development of such long-range coherence, electric waves will sweep either transversely along the cell membrane, and/or perpendicularly towards the centre of the cell.

There are two kinds of collective excitations, according to Fröhlich. One is a metastable state with very high electric polarization, which can arise under very general conditions. The other is coherent electric vibrations, which may be excited by random metabolic energy under more stringent conditions.

The highly polarized metastable state arises because in a polarizable material which mechanically deforms when polarized, the deformation will feed back on the electrical energy. The result is that with increasing polarization, the total energy of the material increases more slowly than it would without deformation. Moreover, this total energy will even decrease with increasing polarization beyond a critical value, and reach a minimum corresponding to a highly polarized state. If this minimum of total energy is negative, the material is ferroelectric and stable against relatively large disturbances. If this minimum is greater than zero, then the state is metastable, and will resist small disturbances only. Now, such polarized states can arise in single protein molecules, or they could be associated with large regions of the cell. Is it possible that polarities, such as the apical–basal of epithelial cells, or the anterior–posterior of the embryo, could initially involve electrically polarized metastable states? This does not preclude other subsequent processes which could stabilize those states, such as ionic currents and expression of particular genes, for example. Ionic currents would tend to reinforce polarization by generating a surrounding field, and so could contribute to stabilizing a metastable polarity in a dynamic way. This polarity may in turn organize proteins in specialized cells such as those of the intestinal epithelium on the basis of their charge characteristics and specific ion-binding capabilities. In the embryo, polarization and ionic currents could be involved in generating and maintaining steep gradients of gene products which have conventionally been credited to the action of passive diffusion or diffusion-reaction (see Driever and Nüsslein Volhard, 1988 for example).

Coherent electric vibrations arise when energy supplied locally to one component is shared by resonant-like transfer to all components vibrating in a similar band of frequencies. When the rate of energy supply exceeds a critical value, the energy will be concentrated in the lowest frequency of the collective. All the components will then vibrate coherently. This mechanism can serve as a very efficient means of storing energy; it can also organize cellular structure and function at macroscopic distances.

Electric polarization waves can co-ordinate the activation of proteins, enzymes, or ion-channels in the cell membrane. Thus, signals impinging on the cell membrane can be readily and quickly amplified. The activation of tranducin by the visual pigment rhodopsin, and the co-ordinated release of Ca^{2+} in the muscle cell during muscle contraction may both occur in this manner. Similarly, the simultaneous firing of many neurons in large regions of the brain associated with various states of brain function and malfunction (see Dichter and Ayala, 1987) may also be co-ordinated by coherent polarization waves which open sodium channels in the cell membrane. (It is known that Ca^{2+}-dependent intracellular enzymic reactions coupled to the gating of ion channels in the cell membrane can generate regular limit cycles of spiking discharges in single cells, see Llinas, 1986. But the precise mechanism for the *coherent* activities of populations of neurons remains poorly understood.)

According to Fröhlich, the oscillations are both transverse and longitudinal. This creates the possibility for coherent vibrations in the membrane to organize intracellular structure and metabolism. The inside of the cell is traversed by a dense network of filaments called the cytoskeleton. The cytoskeleton also connects the cell membrane with the nuclear membrane to which the chromosomes are attached (see Clegg, 1984). In fact, a number of people have suggested that it is more like a system of telegraph wires (see Lomdahl *et al.*, 1984) which enable intracellular communication to occur rapidly via electronic mail, as it were. These wires or cables are made of alpha-helical proteins all capable of transmitting electromechanical vibrations along their lengths. Thus, specific enzymes and substrates can be attracted to these cables at different times in the cell cycle, when the cell is vibrating, perhaps at different frequencies. Different vibrations might even activate the transcription of specific batteries of genes. This could be how metabolism and gene expression is regulated. Del Guidice *et al.* (1986) hypothesize that the vibrations in the membrane can organize and reorganize the cytoskeleton itself, particularly during the gross cellular changes that occur in cellular differentiation or in response to different environments (see next section).

A key question is why coherent vibrations (if they exist) should change in the course of the cell cycle? One possibility is the progressive synthesis of the membrane skeleton (see Bennett, 1985). This is a dense network of fibres directly beneath the cell membrane to which integral membrane proteins are attached. If the membrane skeleton contributes to the mechanical properties of the membrane, then changes in its mechanical properties would alter the frequency of its vibration. Equally, in a spatially heterogeneous cell membrane, it is possible that different sections will vibrate at different frequencies.

To push this line of thought further, one would expect that in a relatively homogeneous membrane, the entire membrane could be involved in globally coherent vibrations. Such globally coherent vibrations could be particularly important in organizing body patterns during early development. As the membrane becomes more and more rigid, due to the synthesis of membrane skeleton, it is expected that higher harmonics of the vibrations will evolve in sequence. Indeed, in *Drosophila* embryos whose

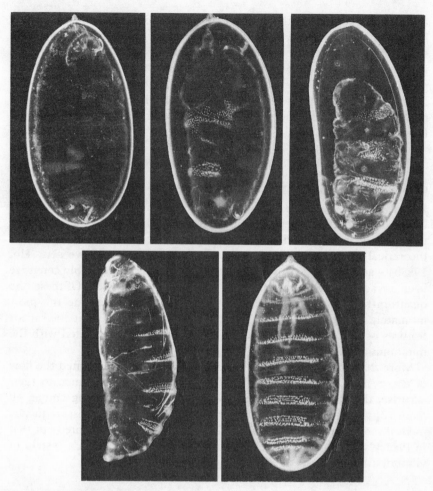

Figure 13.2. Typical transformations in *Drosophila* embryos induced by ether treatment at successively later times which make up a harmonic series from left to right and top to bottom. The last in the series is an embryo with the normal number of segments.

patterning process has been arrested at successively later times by ether-treatment (Ho *et al.*, 1987), one can see the evolution form a simple polarized worm with a tail and very little head, to a body with two divisions, to four, seven or eight, and about sixteen segments in the final fully developed larva (see Figure 13.2). Ether acts by increasing membrane fluidity and ion permeability (Johnson and Bangham, 1969), so its effect may be to inhibit the development of the next higher harmonic or to lead to mode softening. Although the effect of ether is suggestive, the appearance of harmonics *per se* tells us little about the particular mechanism involved, as a lot of different processes could give bifurcations of that kind (for example, classical reaction–diffusion). There are other reasons, however, for suspecting the involvement of coherent excitations in pattern formation. For example, all early animal embryos are characteristically synchronous with regard to nuclei or cell division, indicating that they are, in effect, globally coherent systems. Furthermore, pattern determining processes occur in all animal groups when the embryo is about one mm in length, regardless of the adult size of the animal. This implicates a universal mechanism, such as coherent excitation, that is dependent on absolute linear dimensions.

CONVERGENCE TO A UNIVERSAL ATTRACTOR

It is of interest that several quite independent, theoretical considerations, which I shall outline below, lead to the same prediction of coherent vibrations in a system of oscillators. It suggests that even the most abstract theoretical ideas are rooted in reality – as creative perceptive acts (Ho, 1988b) – and seemingly divergent avenues of thought inevitably converge to an attractor which lies within the structure of reality itself. Of these, the quantum-field theoretic approach (see below) appears to me the most intimately connected to the relevant reality structure, in that the salient features of energy transduction in the living state are predicted with the minimum of *ad hoc* assumptions.

More than thirty years ago, Fermi, Pasta and Ulam calculated the flow of energy in a row of masses connected by springs; and much to their surprise, discovered that the energy, instead of equilibrating among all modes of vibration, became concentrated among the few lowest frequency modes (see Lomdahl *et al.*, 1984). Similarly, Pohl (1980) pointed out that in Planck's formulation for the average energy of a mode in a system of vibrators:

$$ E(v) \ = \ hv\left(\frac{1}{e^{hv/kT} - 1} + \frac{1}{2}\right) $$

one can see that at low frequencies, when $hv/kT \ll 1$,

$$ E(v) \ \approx \ kT $$

and there is equipartition of energy among all modes. At high frequencies, when $hv/kT \gg 1$,

$$ E(v) \ \approx \ hv(e^{-hv/kT}) $$

and there is rapid exponential decay of the average mode energy with increase in frequency. In other words, there is a natural tendency for the oscillations to be concentrated in the lowest frequent modes. Neither of the above considerations, nor Fröhlich's formulation, depend on quantum mechanical assumptions. Yet a number of physicists feel that quantum mechanics is fundamental to the living state.

One recent approach is that of Del Guidice *et al.*, (1986; 1988), who showed that a quantum field theoretic description of the living system as a population, or medium, of electric dipoles, can predict both electric polarization and the spontaneous appearance of coherent, phase-correlated oscillations. The electro-magnetic field (EMF), either endogenously produced, or introduced from an external source, can interact with the coherent medium when their magnitudes are appropriately matched. Because the phase of each oscillator is correlated with the whole medium, it resists change induced by the propagating EMF. The EMF is thereby constricted to propagate along particular pathways where the phase correlation is broken, but is excluded from the bulk medium where correlations are still holding. These pathways form a filamentous network through the medium producing strong field gradients at the boundaries which serve to attract molecules to it. The molecular species attracted will depend on the frequency of the propagating field, giving rise to the possibility of metabolic regulation (see previous section). In this way, the field filaments become matter filaments. The authors suggest that the cytoskeletal network is the embodiment of these electrodynamical effects. Another prediction from the theory is that the effect of external EMFs should not only be frequently specific but intensity-specific. When the EMF is too weak, it fails to propagate through the correlated medium and no effect results. On the contrary, a strong EMF will overcome the correlation of the entire medium so that the effects induced will not be characteristic of the correlated living state.

EVIDENCE FOR COHERENT EXCITATIONS

After the speculations, one might ask what supporting evidence there is that coherent excitations play any role in biological processes. Fröhlich (1980) gave some general predictions concerning processes that involve coherent excitations and are hence capable of being triggered by low-intensity electromagnetic radiation. In the same article, he also reviewed a number of relevant experiments which satisfy those criteria to varying degrees, though most of the findings are disputed by other workers. Nevertheless, new findings continue to be made. McLeod *et al.*, (1987) reported a frequency-specific inhibition of protein synthesis in cultured mammalian fibroblasts subjected to alternating electric fields between 1 hz and 10 hz, which reaches saturation at an amplitude of about 0.1 mV per cm. Moreover, the response is dependent on the orientation of the cells relative to the direction of the applied electric field. In a series of well designed experiments, Chan and Nicholson (1986), using much higher field-strengths (which were still orders of magnitude below cellular membrane potentials), showed that the firing patterns of both Purkinje cells and stellate interneurones were modulated in a manner that depended on their dentritic orientation with respect to the field.

The most convincing biological effects were those obtained with very low frequency electromagnetic radiation. Frequencies of about 10 hz and below match certain intrinsic coherent oscillations in firing patterns of neurons that can occur in the central nervous system (see Llinas, 1986), and this may constitute the best evidence by far for coherent excitations. Do higher frequencies exist as coherent vibrations in living systems?

Popp (1986; 1988) and co-workers observed coherent weak photon emission from cells and embryos within the visible range, which are characteristic of physiological state and developmental stage. He views the organism as a 'multimode laser' generating coherences at a wide range of frequencies, which may be released as electromagnetic radiations or visible light, so long-range coherence could be communicated within and between organisms.

In addition to the experimental observations, there have been persistent claims linking overhead power lines and cases of cancer and leukaemia in children living near them, so much so that the Central Electricity Generating Board in Britain, which has resisted those claims for many years, is now organising its own investigation (see article in *The Times*, Friday March 18, 1988).

Despite the observations, there is a general feeling of dissatisfaction with the data. Part of the difficulty is that in order to investigate coherent excitations we are required to study living organization as it is living and developing. Our traditional methodological and conceptual framework on the whole is not yet equal to the task. The application of novel non-invasive technologies, such as the SQUID magnetometer, nuclear-magnetic resonance imaging, acoustic microscopy, and so on, should overcome some of the methodological difficulties, *if* we know what we should be looking for.

Many of the experiments that have been done give only indirect information concerning coherent excitation in that they depend on all-or-none effects of electromagnetic radiation at particular frequencies. The living system is essentially acting as a super-sensitive sensing device, as almost no physical instrumentation to date can match that densitivity. Given that the effects are developmental-stage specific, then a positive or negative result will be obtained depending on whether the population of experimental material is sufficiently synchronised. The more satisfactory approach is to demonstrate coherent excitations directly, rather as Popp and his colleagues are doing. Here, synchronization remains a problem particularly if the emissions are weak to begin with. Ideally, one should aim to obtain data from single developing organisms in which the developmental stage and electrical activities can be monitored directly in the presence and absence of EMF. I have begun working on such a system for *Drosophila* embryos in collaboration with Charles Nicholson. The preliminary results already give an inkling of the exquisite light and sound concert that is life itself.

REFERENCES

Arnott, C. and McDowell, C.A. (1958) The quenching of the iodine fluorescence spectrum. *Can. J. Chem. 32*, 114–20.

Bennett, V. (1985) The membrane skeleton of human erythrocytes and its implications for more complex cells. *Ann. Rev. Biochem. 54*, 273–304.

Chan, C.Y. and Nicholson, C. (1986) Modulation by applied electric fields of Purkinje and stellate cell activity in the isolated turtle cerebellum. *J. Physiol. 371*, 89–114.

Clegg, J.S. (1984) Properties and metabolism of the aqueous cytoplasm and its boundaries. *Am. J. Physiol. 246*, R133–R151.

Del Guidice, E., Doglia, S. and Milani, M. (1986) Spontaneously broken symmetries and dissipative structures. In (C.W. Kilmister, (ed.) *Disequilibrium and Self-Organization*. Dordrecht: D. Reidel Publishing Company.

—— (198) Structure, correlation and electromagnetic interaction in living matter. Paper presented at *Workshop Conference on Basic Issues in the Overlap and Union of Quantum Theory, Biology and the Philosophy of Cognition*. Bermuda, April 15–27.

Dichter, M.A. and Ayala, G.F. (1987) Cellular mechanisms of epilepsy: a status report. *Science 237*, 157–64.

Driever, W. and Nüsslein-Volhard, C. (1988) A gradient of *bicoid* protein in *Drosophila* embryos. *Cell 54*, 83–93.

Fleming, G.R., Martin, J.L. and Breton, J. (1988) Rates of primary electron transfer in photosynthetic reaction centres and their mechanistic implications. *Nature 333*, 190–2.

Frohlich, H. (1968) Long range coherence and energy storage in biological systems. *Int. J. Quantum Chem 2*, 641–9.

—— (1980) The biological effects of microwaves and related questions. *Adv. Electronics and Electron. Phys. 53*, 85–152.

Hibbard, M.G., Dantzig, J.A., Trentham, D.R. and Goldman, V.E. (1985) Phosphate release and force generation in skeletal muscle fibres. *Science 228*, 1317–19.

Ho, M.-W. (1988a) On not holding nature still: evolution by process not by consequence. In M.-W. Ho and S.W. Fox (eds), *Evolutionary Processes and Metaphors*. London: Wiley.

—— (1988b) Re-animating nature: the integration of science with human experience. *Public lecture for Scientific and Medical Network*, May 21, 1988.

Ho, M.-W., Matheson, A., Saunders, P.T., Goodwin, B.C. and Smallcombe, A. (1987) Ether-induced segmentation disturbances in *Drosophila melanogaster*. *Roux's Arch. Devel. Biol. 196*, 511–21.

Huxley, H.E. (1971) The structural basis of muscle contraction. *Proc. Roy. Soc. London B, 178*, 131–149.

Llinas, R.R. (1986) Neuronal oscillators in mammalian brain. In M.J. Cohen and F. Strumwasser (eds), *Comparative Neurobiology: Modes of Communication in the Nervous System*, pp. 279–290. New York: Wiley & Sons.

Lomdahl, P.S., Layner, S.P., and Bigio, I.J. (1984) Solitons in biology. *Los Alamos Science*, Spring 1984, 3–22.

McClare, C.W.F. (1971) Chemical machines, Maxwell's demon and living organisms. *J. Theor. Biol. 30*, 1–34.

—— (1972) A 'molecular energy' muscle model. *J. Theor. Biol. 35*, 569–5.

McCleod, K.J., Lee, R.C. and Ehrlich, H.P. (1987) Frequency dependence of electric field modulation of fibroblast protein synthesis. *Science 236*, 1465–8.

Popp, F.-A. (1986) On the coherence of ultraweak photoemission from living tissues. In C.W. Kilminster (ed.), *Disequilibrium and Self-Organization.* pp. 207–230. Dordrecht: Reidel.

—— (1988) Experiments on the creation of weak coherent biophotons by the DNA-molecule. Paper presented at *Workshop Conference on Basic Issues in the Overlap and Union of Quantum Theory, Biology and the Philosophy of Cognition.* Bermuda, April 15–27.

Stryer, L. (1987) The molecules of visual excitation. *Scientific American 257*, 42–50.

14. Discrete aspects of morphogenesis and gene dynamics

G. COCHO and J.L. RIUS

More than a century has passed since the publication of Darwin's *On the Origin of Species* and more than forty years since the 1947 Princeton congress where the synthetic theory of evolution was established. Although this synthetic theory has helped to understand many features in different fields (genetics, taxonomy, paleontology, etc.), it is in some way rather loose and descriptive, instead of being more restrictive and analytical. That is why, in the last two decades, many scientists have argued that structural and dynamical constraints due to the physical and chemical characteristics of living organisms might restrict in an important way the structural and dynamical spectrum of living systems. We believe that was perhaps the main driving force behind the four Waddington conferences: 'Towards a theoretical biology'.

In many of these discussions and attempts, the analytical (or 'physicalist') approach has been often associated with what one might call an ahistorical approach as a counterpart to the historical aspects. However, in recent years, the study of disordered materials with long-range conflicting interactions (e.g. spin glasses, see Kinzel, 1987) has shown that the state of these materials depends on the previous history and that they indeed have memory and cognitive potential. Spin glasses have a discrete structure and their constituents (atoms or molecules) have long-range interactions which may be attractive or repulsive in a more-or-less random way. On the other hand, taking into account the discrete aspects of multicellular organisms (which are built of cells) and taking into account the short range cell–cell interactions, the colour patterns of different animals have been analysed and classified (Cocho *et al.* 1987a, b).

In this paper we will try to show that models based on the discrete aspects of living organisms have the capacity to explain some of the analytical and historical features of morphogenesis and evolution. It would seem that the modern-physics approach might help to arrive at a synthesis of some of the analytical and historical aspects of living systems and we might talk of a 'physicalist synthetic theory of evolution'.

We will now discuss constraints associated with discrete aspects of living organisms, at both the cellular and molecular level and for both short- and long-range interactions.

SHORT-RANGE INTERACTIONS

Local energy density and gene dynamics

The synthesis of oligonucleotides has allowed rather precise estimations of the enthalpy, entropy and free energy of nearly base pairs ('digrams') for both RNA (Freier *et al.*, 1986) and DNA duplex (Breslauer *et al.*, 1986). Table 14.1 shows that for RNA, both the enthalpy, ΔH, and the free energy, ΔG, show similar digrammatic dependence. If we label adenine (A) and uracil (U) bases by W (weak), and guanine (G) and cytosine (C) by strong (S) we can divide the data of Table 14.1 into three regions corres-

Table 14.1. Enthalpy ΔH and free energy ΔG for RNA and DNA digrams.

	RNA		DNA	
Digrams	ΔH	ΔG	ΔH	ΔG
AA	6.6	0.9	9.1	1.9
AT	5.7	0.9	8.6	1.5
TA	8.1	1.1	6.0	0.9
TT	6.6	0.9	9.1	1.9
CC	12.2	2.9	11.0	3.1
CG	8.0	2.0	11.9	3.6
GC	14.2	3.4	11.1	3.1
GG	12.2	2.9	11.0	3.1
AC	10.2	2.1	6.5	1.3
AG	7.6	1.7	7.8	1.6
TC	13.3	2.3	5.6	1.6
TG	10.5	1.8	5.8	1.9
CA	10.5	1.8	5.8	1.9
CT	7.6	1.7	7.8	1.6
GA	13.3	2.3	5.6	1.6
GT	10.2	2.1	6.5	1.3

Note: ΔH and ΔG are in Kcal/mol

Table 14.2. Average values for 23 sequences

		xy	yz	zx
ΔH	RNA	9.52	9.99	10.66
ΔG	RNA	1.92	2.02	2.25
	k^1	0.82	0.58	0.32
	k^2	0.65	1.82	0.86
	k^3	1.53	0.59	1.12
	k^4	1.00	1.01	1.70

ponding to W–W, W–S plus S–W and S–S. Different values of these duplex digrammatic energies might imply different axial and torsional flexibilities and one may wonder if, in the ribosome tRNA–mRNA duplexing process, different codon links (x–y, y–z and z–x) have different averaged enthalpy and/or free energy. The averaged values for twenty-three exonic sequences (eighteen from vertebrates and five from *Drosophila*) are shown in the first two rows of Table 14.2. It is found that $\Delta Hxy < \Delta Hyz < \Delta Hzx$ and the same comment is valid for ΔG and for all twenty-three exons, $\Delta Hzx > \Delta Hxy$ and $\Delta Gzx > \Delta Gxy$. Hence, it seems that in the duplexing translation process the x–y is, on average, weaker and more flexible than the z–x one, with y–z having, in general, intermediate values.

One can define

$$k^1 = (N_{AA} + N_{AU} + N_{UA} + N_{UU}/\textstyle\sum$$
$$k^2 = (N_{AC} + N_{AG} + N_{UC} + N_{UG}/\textstyle\sum$$
$$k^3 = (N_{CA} + N_{CU} + N_{GA} + N_{GU}/\textstyle\sum$$
$$k^4 = (N_{CC} + N_{CG} + N_{GC} + N_{GG}/\textstyle\sum$$
$$\textstyle\sum = (k^1 + k^2 + k^3 + k^4)/4$$

where N_{ij} is the frequency of the digram ij. Random sequences with equal probability for the four bases should have k^i values near to one. One would expect W–W digrams (k^1) to be more abundant in the x–y links and S–S digrams (k^4) to dominate in z–x. However, it is not possible to fulfill both conditions in the z–x–y 'trigram' and lower energy allowed trigrams will be S–S–W and S–W–W, implying that W–S digrams should dominate in y–z links. We have analysed the average values of the four k's, for the three types of links in the twenty-three exonic sequences. These values are shown in the last four rows of Table 14.2 where k^1 dominates in x–y, k^2 in y–z and k^4 in z–x. For all the twenty-three exonic sequences $k^1_{xy} > k^1_{zx}$, $k^4_{xy} < k^4_{zx}$, $k^2_{xy} < k^2_{yz}$ and $k^3_{xy} > k^3_{yz}$. These *RNA* periodicities imply non-random aspects in the genetic code. As can be seen from Figure 14.1, in most of the twenty-three exonic sequences, S bases dominate at the z position and, in order to break the z degeneracy, the most natural way is to distinguish between purines and pyrimidines (true in the genetic code). One might also argue that as the RNA periodicities act also at the x and y positions, the discrimination between important amino acid properties might be correlated with a purine–pyrimidine dichotomy. Again, this statement seems to be true. If one divides the amino acids codons into two groups, one of them with a purine at the y position and the other one with a pyrimidine, one finds:

Purine-y: lys, asn, gln, his, glu, asp, tyr, arg, ser, gly, trp, cys.

Pyrimidine-y: thr, pro, ala, ser, ile, met, leu, val, phe.

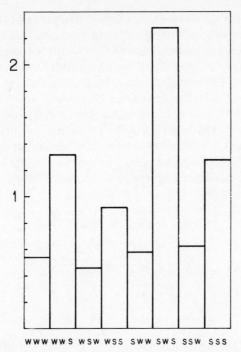

WWW WWS WSW WSS SWW SWS SSW SSS

Figure 14.1. Codon frequencies, in the weak-strong degenerate representation, for the 23 exonic sequences. A random sample should show all values near to one.

Almost all purine–y amino acids are polar and almost all pyrimidine–y ones are non-polar (serine is represented in both groups and cysteine, non-polar, is in the purine–y group). Note that the classification in polar and non-polar is quite different from the division into hydrophilic and hydrophobic. Although the preceding comments do not give any clue to these purine–polar and pyrimidine–non-polar correlations, they indeed suggest a primitive scenario where water was less important than now.

In the preceding analysis, one knew from the beginning that the twenty-three sequences were exons. Local Fourier transforms for the four k^i's allow one to distinguish exons from introns. Let $k^i(n)$ (with $i = 1$ to 4 and $n = 1$ to $N - 1$) be the values of the quantity k^i for the digrams of a given sequence of length N. We define the power spectrum $P_j^i(f)$ (at the point j) by:

$$P_j^i(f) = [C_j^i(f)^2 + S_j^i(f)^2]/L$$

$$\text{where} \quad C_j^i(f) = \sum_{r=j-l/2}^{r=j+L/2} k^i(r) \cos (2\pi rf/L)$$

$$\text{and} \quad S_j^i(f) = \sum_{r=j-l/2}^{r=j+L/2} k^i(r) \sin (2\pi rf/L)$$

f is the frequency and L the width of a window centred at the digram j. One is interested in the frequency $f3 = \text{Int}(L/3)$, the frequency associated to a period of three bases. The function $P^i_j(f3)$ is a 'semi-local measure' of the intensity of the three-base periodicity of the quantity k^i around the j digram. These semi-local values does not depend on the frame phase, which might be different in different regions. One would expect $P^i_j(f3)$ to be important in exonic regions and small in intronic ones. Although smaller $P^i_j(f3)$ should be important in pseudogenes or 'intronized exonic regions' with a small number of insertions or deletions. These 'intronized exons' would lose more or less slowly the three-periodicity. Power spectra where the frequency $f3 \pm 1$ were considered might localise the intronized exons and perhaps estimate their age.

The $f3$ and $f3 \pm 1$ for human β-globin and *Drosophila melanogaster* ADH are shown in Figure 14.2. We have taken $L = 120$ and sampled every five bases ($j = 5\,s$ with s an integer). The exonic and intronic regions are well distinguished, although at the beginning of the second intron and in the five-region of human β-globin, regions with exonic features are found. In these two regions the power spectrum for $f3 \pm 1$ is higher than for $f3$, suggesting that the two exon-like regions have 'aged'.

Table 14.1 shows that for DNA the binding between base pairs of equal 'strength' (considering A and T as weak and G and C as strong) is larger than the binding between pairs of different strength. One may argue that low local DNA values of ΔH and ΔG would imply larger flexibility and fluctuations. These larger fluctuations might be associated with a larger frequency of duplication errors and point-mutations would be more frequent at the weak w digrams (W–S and S–W) than at the strong s ones (W–W and S–S).

We may compare point-mutations where there is no change in the two neighbouring bases and limit ourselves to the C–G and T–A transversions where the strong or weak character does not change and the direction of the mutation does not matter. As can be seen in Table 14.1, CC, CG, GC, GG, AA, AT and TT digrams are stronger (S) than the remaining ones. In Table 14.3 we show the numbers for W—W, W–S, S–W and S–S 'trigrams' for four sequences. The total number are: $N_{W-W} = 24$, $N_{W-S} = 16$, $N_{S-W} = 2$ and $N_{S-S} = 11$. The C–G and T–A transversions are more frequent in weak links and in particular when the first line of the trigram is a weak one. Although better statistics are needed, it seems that there exist correlations between the value of the local enthalpy and the frequency of point-mutations. This feature and the RNA periodicity restrictions discussed previously would interact with the 'external restrictions' associated with natural selection, modifying the predictions of neutral or other theories.

DISCRETE CELL–CELL INTERACTIONS AND PATTERN FORMATION

There are basic concepts:

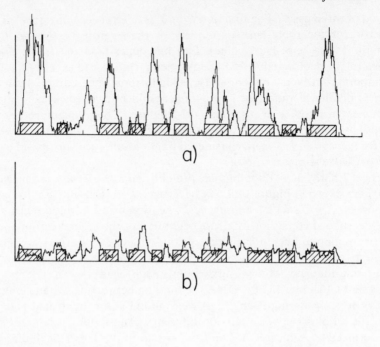

a)

b)

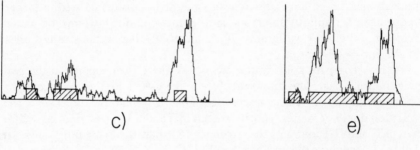

c)

e)

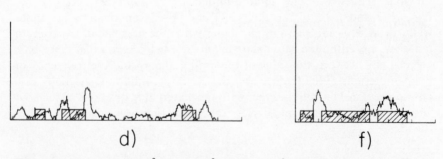

d)

f)

Figure 14.2. Power spectra P_j^2 (f3) and $[P_j^2$ (f3 $-$ 1) $+ P_j^2$ (f3 $+$ 1)]/2 for human steroid-hydroxylase [(a) and (b)], human β-globin [(c) and (d)] and *Drosophila* alcohol – dehydroxylase [(e) and (f)]. Dashed lines represent the exonic sequence regions. Note the exonic structure, at the beginning of the second intron and in the 3′ region of human β-globin. For both structures, the ratio $[P_j^2$ (f3 $-$ 1) $+ P_j^2$ (f3 $+$ 1)]/2P_j^2 (f3) is quite large, suggesting that they are somewhat old.

Table 14.3. Trigram frequencies in C–G and T–A transversions

Sequences	N_{W-W}	N_{W-S}	N_{S-W}	N_{S-S}
Plasmodium Falciparum	6	1	0	1
Plasmodium Falciparum	6	1	0	1
Glob–Mouse α/β	6	3	1	2
Apolipo. E Human/Rat	6	11	1	7
Total	24	16	2	11

1) All living organisms have discrete aspects (i.e. they are built from cells).
2) For a given tissue or organ, the number of different cell types is rather small (we may consider a 'basic cell' in a small number of states).
3) Cells can interact among themselves by contact (e.g. cell adhesion molecules CAMs.) or short-range (e.g. morphogens) interactions (in analogy with physics of materials, where atoms or molecules in lattices interact among themselves).
4) Cells can have a non-systematic, random-like motion (such as thermal motion in thermodynamics).

These four aspects suggest an equilibrium statistical approach to cell dynamics.

5) In development, cells may freeze sequentially in rows or sheets (as in directional solidification of metals, out of equilibrium features which can be modelled with cellular automata).
6) Division and multiplication of cells including clonal growing (non-polynomial 'cell division automata').

The relative importance of the different mechanisms and the sequential or simultaneous operation of these mechanisms will imply different scenarios. We will consider two scenarios:

Scenario 1. The sequential action of:
1) The accommodation (minimizing an 'energy') of a rather small number of cells (statistical mechanics).
2) The sequential growing and freezing (unidirectional solidification of metals) modeied by cellular automata.
3) The clonal growing of the frozen cells.

We will present the colour patterns of snakes and butterflies as examples.

Scenario 2. The alternation of division and 'minimal energy' accomodation (as the cells or nuclei multiply the density may increase and the effective cell–cell or nucleus–nucleus interactions might change).

We will present a model of the formation of *Drosophila* egg 'homeotic bands'.

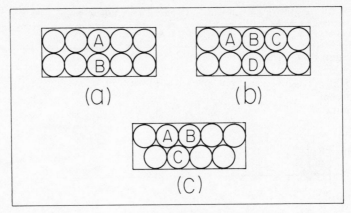

Figure 14.3. Cellular automata: (*a*) S20 (square lattice, two states, zero-range neighbour). The state of the element B depends only on the state of element A. (*b*) S21 (the state of D depends on the state of A, B, and C. (*c*) T21 (triangular lattice, two-state, first-range neighbours). The state of C depends on the states of A and B.

A MODEL FOR THE MORPHOGENESIS OF COLOUR PATTERNS IN SNAKES

Cellular automata are systems built from discrete elements which can be in a finite number of states. These elements change in discrete time steps following simple rules. For two-dimensional growth of crystals or biological systems we can use square or triangular lattices. One can classify these automata by the type of lattice (square, S, or triangular, T), the number of states (k) allowed to an element and by r (associated to the number of precursor elements which define the state of the successor).

If one considers that contact (CAM-like) cell–cell interactions dominate, only $r = 0$ and 1 cellular automata will be relevant (Figure 14.3).

The morphogenesis of colour patterns in snakes can be modelled from the following assumptions:

1) At the dorsal region of the snake embryo a row of a rather small number of cells interact through short-range interactions, and simple, periodic configurations minimise an 'energy'.

2) This periodic row is used as the initial condition for cellular automata which model the sequential division and freezing. If one assumes that, due to the dominance of contact interactions, only cellular automata with r = 0 and 1 are relevant, one can classify the possible dynamical rules and hence the possible patterns. One finds most of them in animals and all the colour patterns of snakes can be built from a subset of these rules (Figure 14.4). The colour pattern of large felines and some fishes can also be obtained.

Butterflies

Most of the colour patterns of butterflies can be derived from the superposition of elementary patterns where the 'original seeds' for the action of

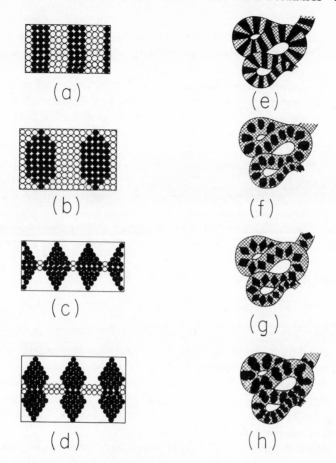

Figure 14.4. Cellular automata and colour pattern of snakes. The initial row is in the center of the figure and grows upwards and downwards. (*a*) S20 or S21, rule 1100. (*b*) S21, transition from rule 1100 to 1000. (*c*) T21, rule 100. (*d*) T21, transition from rule 100 to 110. Colour pattern of snakes: (*e*) Banded krait (*Bungarus fasciatus*). (*f*) Prairie rattlesnake (*Crotalus v. vividis*). (*g*) Mexican west-coast rattlesnake (*Crotalus basilicus*). (*h*) Sarae bintade (*Bothrops newiedi*).

the cellular automata rules are a dot, a short row and a line of dots (see Figure 14.5). Distortions, displacements and unequal growth (associated to boundary conditions) may modify these elementary patterns.

A MODEL FOR THE FORMATION OF THE 'HOMEOTIC BANDS' IN DROSOPHILA EMBRYO

1) The 'homeotic bands' appear when the nuclei, after dividing *n*-times, are at the surface.

2) At first, one observes 'isolated dots', then a 'homogeneous covering', and finally, the covering breaks into transversal 'zebra' bands.

3) The bands have been associated to the 'passive response' of the nuclei to standing reaction–diffusion waves at the egg surface.

4) In variance with 3) we will assume that a dominant mechanism is the interaction among nuclei when the 'nuclear density' is large enough. If the 'internuclear distance' is large, there will not be significant nucleus–nucleus interaction and only self-replication or self-inhibition will be allowed. That would be the case for the 'isolated dots' and 'homogeneous covering' stages. If the average distance among nuclei is small enough, the nucleus–nucleus interaction will be relevant and bands may form if inhibitory interactions are present. The transversal character of these bands might be associated with the influence of the posterior polar cells. It is worth remarking that local self-enhancing and competitive inhibitory interactions have already been discussed in the literature; it seems, therefore, that experimental evidence exists to support the mechanism of the model. However, other (perhaps more important) mechanisms might be present.

The model

1) We assume that the nuclei can be in two states: segregating or not segregating 'homeotic products'.

2) The self-enhancing nucleus can be modelled by an external field.

3) The nucleus–nucleus inhibition is modelled by an antiferromagnetic interaction (in a periodic square lattice). For a given surface nuclear density we minimise 'an energy'.

4) It is assumed that the anterior and posterior regions are 'black' (no homeotic products). These boundary conditions imply the formation of transversal, instead of longitudinal, bands.

5) After a given stage (cellular blastoderm) the pattern will be frozen.

As an example, let us consider a rectangular domain, eight lattice points wide and sixteen lattice points high (more or less the relative dimensions for a cut and flattened egg surface). The forward and rear sites are assumed to be black and the lower borders obey periodic boundary conditions (closing

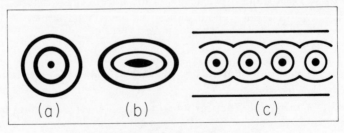

Figure 14.5. Butterfly elementary patterns.

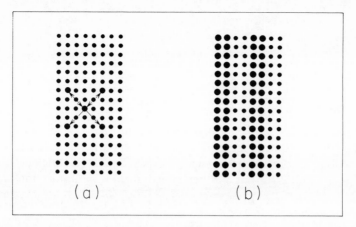

Figure 14.6. (*a*) Square lattice and inhibitory 'antiferromagnetic' diagonal interaction. (*b*) Minimal energy pattern for the lattice and interaction shown in (*a*).

the cylinder). We then consider diagonal antiferromagnetic interactions (see Figure 14.6a). The minimal energy configuration is shown in Figure 14.6b. Note that all points have four antiferromagnetic neighbours at the diagonal interactions distance. Montecarlo calculations show that the pattern is stable if other weaker interactions are added.

It is worth remarking that for 'diagonal antiferromagnetic' interactions the covering $c = 1/2$ (c = number of blacks divided by the number of blacks plus whites) minimises the energy and that for 'diagonal' antiferromagnetic interactions at a diagonal distance, n, the bands have width $w = n$. For $n =$ odd, there exists another solution with the same minimal energy: equal spaced bands of width $w = 1$. However, a weak short-range interaction will break the degeneracy favouring the $w = n$ bands. The presence of an autocatalytic self-interaction modelled by an external field modifies the value of the covering c and alternated bands of different width can be present.

LONG-RANGE INTERACTIONS

The study of disordered materials with long-range interactions, random in sign (e.g. spin glasses) has shown rather interesting features. The phenomenological characterization of the spin-glass phase comprises the presence of hysteresis cycles, remnant magnetization dependent on the past history of the sample and long relaxation times. All these effects suggest the existence of many stable or at least metastable equilibrium states. In the spin-glass energy landscape it is possible to define an ultrametric distance between two local minima by minimal energy barrier separating them.

Given three points, A, B and C, a metric space is ultrametric if the inequality $d(A,C) < \text{Max}\ (d(A,B), d(B,C))$ is satisfied.

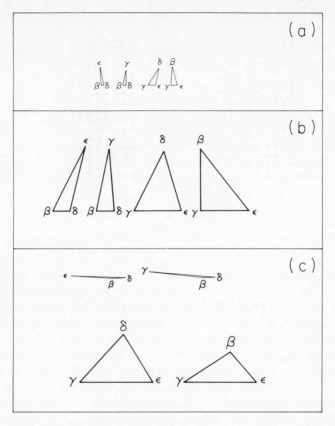

Figure 14.7. Hamming distance triangles for some of the data of Table 13.4. (*a*) second exon replacement sites. (*b*) second shown in exon silent sites. (*c*) 5′ non-coding region.

This inequality implies that the three points form an equilateral or isosceles triangle (with one of the sides smaller or equal to the other two equal sides). This ultrametric property can be used as the basis of hierarchical taxonomical trees where the distance between two entities is given by the distance to the nearest common node.

Ultrametricity emerges as a consequence of randomness and the law of large numbers, the presence of isosceles ultrametric triangles in spin glasses and many optimization problems are associated to the valley structure of the energy landscape of these disordered frustrated systems. By frustration (Toulouse) one understands the impossibility of fulfilling simultaneously the constraints associated to different conflicting interactions.

Due to these 'historical properties', dynamical models based on some of the features of spin glasses have been applied to neural networks, protein structure and evolution. As noted at the beginning of this paper, it seems that discrete systems with short- and long-range conflicting dynamics have the capacity of taking into account both 'analytical and historical' features, helping to establish what might be called a 'physicalist synthetic theory of evolution'.

It is worth noting that in the dynamics of DNA point mutations, both short- and long-range constraints seem to be present. A short-range constraint is the non-random point mutation component (for both introns and exons) discussed previously. Also the x–y–z RNA periodicities. Long-range constraints would be associated with amino acid–amino acid interactions in proteins. The objects of these interactions are separated along the α-helix and β-regions and rather distant, along the polypeptide chain, when the protein folds. These long-range interactions should imply long-range correlations along the puric and pyrimidic bases of exonic DNA. Due to these long-range random-like constraints, one would expect ultrametric features in the point-mutations (and therefore, the existence of taxonomic trees). Such a behaviour might not be true for introns and intergenic regions due to the exclusive presence of short-range constraints (however, long-range constraints associated to the superhelix and nucleosome structure might be present).

As a matter of fact such behaviour is found in globins. Maniatis *et al.* in a series of articles have carried out an extensive analysis of the human β-globin family, including comparison with other mammals. They have computed the divergences of the coding sequences (replacement sites and silent sites), 5′ flanking regions, 5′ non-coding regions and 3′ non-coding regions. Some of their results are presented in Table 14.4. From these data one can compute the metric triangles of Figure 14.7. The length of the sides of the triangle are given by the corrected Hamming distance of Table 14.4.

Figure 14.7 shows that, for the exonic sequences the silent-site triangles are less ultrametric than the replacement-site ones. The 5′ and 3′ triangles are even less ultrametric. Therefore, silent sites and non coding-sequences (which from a random-like neutralist dynamics one should observe ultrametrical character), show less ultrametrical features than the replacement sites, where the non-local constraints associated with the translation of functional properties are important. It seems to us that the data are

Table 14.4. Corrected percent fivergences of human β-globin gene family.

Gene pair	5′ flanking region	5′ non coding region	2nd exon r.s.	2nd exon s.s.	3′ non coding region
β/δ	34	8.5	2.2	16	59
γ/ϵ	43	65	6.1	43	71
β/γ	45	49	15	54	68
δ/γ	63	57	14	53	86
β/ϵ	56	35	16	66	57
δ/ϵ	72	49	17	59	65

Note: r.s. = replacement sites. s.s. = silent sites. The second exon is the longer one and code a protein domain which binds to the heme.

consistent with a dynamics where, in the absence of these non-local constraints, non-random local dynamical aspects are responsible for the non-ultrametric behaviour. The ultrametric character of the mutations in displacement sites might be due to the long-range spin-glass-like correlations associated with the translation in functional proteins. The presence of interrelations between neutral random mutations and short- and long-range constraints might be the basis of a realistic modelling of many of the features of mutation dynamics. In morphogenetical dynamics it is possible that the more or less conflicting influence among the mechanisms associated with short-range cell–cell interactions and long-range interaction due to chemical diffusion might be the basis of a successful modelling of the morphogenesis of multicellular organisms.

REFERENCES

Breslauer, K.J. *et al.* (1986) Predicting DNA duplex stability from the base sequence. *Proc. Natl. Acad. Sci. 83*, 3746–50.

Cocho, G., Pérez-Pascual, R. and Rius, J.L. (1987a) Discrete systems, cell–cell interactions and color pattern of animals. I. Conflicting dynamics and pattern formation. *J. Theor. Biol. 125*, 419–35.

Cocho, G., Pérez-Pascual, R., Rius, J.L. and Soto, F. (1987b) Discrete systems, cell–cell interactions and color pattern of animals. II. Clonal theory and cellular automata. *J. Theor. Biol. 125*, 437–47.

Efstratiadis, A. *et al.* (1980) The structure and evolution of the human β-globin gene family. *Cell 21*, 653–68.

Freier, S.M. *et al.* (1986) Improved free-energy parameters for predictions of RNA duplex stability. *Proc. Natl. Acad. Sci. 83*, 9373–77.

Kinzel, W. (1987) Spin glasses and memory. *Physica Scripta 35*, 398–401.

15. Time and space scales in neurophysiology

A.A. MINZONI and F. ALONSO-deFLORIDA

The purpose of this paper is to show how well-established ideas in continuum mechanics can be used to understand physiological phenomena. We feel that some physiological systems can be examined by applying these ideas, not only as a qualitative approach, but also to make precise statements about measurable physiological parameters regarding time and space scales. The formulation of phenomena occurring on separated scales allow us to understand the emergence of properties from the micro- to the macroscopic level of organization. As continuum mechanics, physiology is concerned with the average behaviour of macroscopic structures.

We begin by describing the analogy between neurophysiology and continuum mechanics, and then go on to describe some examples of our research where the ideas of fluid and solid mechanics seem to have direct counterparts in neurophysiology. We shall also mention some solutions and possible ways to approach other unsolved problems. Interestingly enough, these problems have not been solved, either in continuum mechanics or in physiology. There is a short appendix which gives a schematic presentation of the mathematics used in the examples and a proposal for one of the unsolved questions.

TYPES OF MODELS FOR EMERGENT SCALES

The philosophical basis of our approach stems from the fact that phenomena in general are presented in levels of order. Entities at one level interact as parts which constitute wholes of a higher level of systematization; and these in turn may interact and unite to form wholes of still higher levels. This idea of levels of order, almost completely neglected in past theoretical research in physiology, is currently a matter of increasing interest. However, we know that explanation of a given phenomenon requires analyses (i.e. it requires understanding of the interaction of its parts); explanation can therefore be defined as a representation of how some properties in an entity emerge from the behaviour of its parts. However, explanation need not to go all the way down through levels. Rather, the understanding of the given phenomenon does not require an explanation of its parts, but only their description. Thus, physiology may be concerned in some instances with the behaviour of large systems of cells without being much interested in the biochemical processes involved in each individual cell. The physiologist need not go all the way upwards either. For instance, he may be interested in neural nets but would not necessarily attempt to answer questions in the realms of ethology or psychology. Similarly, in the non-living realm, continuum mechanics is concerned with questions related to the behaviour of fluids and solids as entities arising from the collision of molecules, without paying much attention to the analyses of the individual behaviour of these molecules. The upper cut-off does not arise in physics, because fluids and solids are simpler systems and the question of upper behaviour as an emergent property does not arise.

On these grounds, even if there are large differences in complexity between the aims of physiology and those of continuum mechanics, both show a sharp demarcation between two levels of order which can be equally formulated as corresponding to different time and space scales. Therefore, the ideas and techniques developed to explain a macroscopic physical phenomenon in terms of microscopic ones can be used to gain insight into the levels of the biological process.

First, an important model borrowed from continuum mechanics which can be applied to physiology is the one of *modulation* (Kevorkian and Cole, 1981). In fluid systems there are structures (phenomena), such as plane waves, periodic oscillations, vortex systems, etc., which can be excited by various initial conditions. The initial and boundary conditions are measured by characteristic length and time scales. At the start, excitation involves both the intrinsic interactions in one scale and the interaction with the outside via the initial and boundary conditions in another. When the scales differ to a considerable extent, the developed phenomenon can be well described by a modulation of the elementary solutions. The explanation then involves the emergence of a modulated phenomenon in terms of the average events occurring in a small scale.

Second, there is the idea of *exchange of stability*, or bifurcation of solutions (Gukenheimer and Holmes, 1983). This process is characterized as the emergence of a phenomenon occurring in a large and slow scale due to the transfer of energy from an unstable state to macroscopic perturbations. Again, this process can be described as a modulation involving the coupling of the lower state with an unstable structure. From a mathematical point of view, the average behaviour of the system is described as an interaction of the elementary solutions with the appropriate modulations imposed upon them.

Third, there is the notion that new phenomena can develop as interactions of multiscale structures, together with scales imposed by nonhomogeneities in the system (Gukenheimer and Holmes, 1983). Again, this situation can be described in terms of appropriately modulated structures.

THE KINDLING EFFECT AS AN EXAMPLE OF MODULATION

A major neurophysiological problem is to explain so-called *plasticity* (Matthies, 1982). This shows how very marked patterns of nervous activity emerge in a level or neural organization in a slow and large scale, evolving from the activity and interaction of elements of the immediately lower level, where the events occur in a fast and small scale. Both duration and lengths are important in plastic neural changes.

The mathematics so far used to explain the behaviour of fluids and solids in physics were applied in an attempt to understand the kindling effect which is a laboratory phenomenon exhibiting plasticity (Goddard and Douglas, 1976). To produce the kindling effect, electrodes are implanted into the brain of rats, cats or monkeys. Special sites of the limbic system are chosen. Once recovered from the operation, the animals can continue their normal lives in spite of the permanently implanted electrodes. Very low intensity pulses (less than 1 mA) at low frequency (from 60 to 100 Hz) and short duration (about 1 s) are then administered periodically (at 12 to 24 h periods) for 7–14 weeks. In the initial days after such small perturbation, there is no observable drastic change in the behaviour of the animals when the stimulus is applied. However, some weeks thereafter, the phenomenon matures and becomes characterized by the appearance of a very dramatic generalized convulsion with loss of consciousness, which is elicited as a response to the invariant train of stimuli periodically administered through the previous weeks. If one looks at the neuronal activity by recording the electric changes produced, the evolution of the phenomenon can be ascertained. At the start, superimposed on a background oscillation, the train of stimuli elicits a monotonic response. During the following days this response evolves, first into a damped oscillation, and later into an afterdischarge (i.e. a self-sustained oscillation of uniform amplitude, characteristic frequency and definite duration). The phenomenon is produced after a lag or latency measurable from the time the train of stimulating pulses comes to an end. In subsequent days, the iterated stimulation causes the afterdischarge to evolve further, thus exhibiting progressive changes in amplitude and latency. Interestingly enough, it becomes progressively more complex, being composed of an increasing number of basic frequencies until the convulsion in finally produced. This kindling effect evolves not only in time, but also propagates in space. Initially, it is confined to the directly stimulated nuclei, later spreading slowly (on a scale of weeks) to encompass successively the elements of a chain of neural modules (or networks) along the natural paths of the projecting axons.

The kindling effect appears as a very complex phenomenon, but we attempted to approach the explanation of certain aspects of it (Alonso-deFlorida and Minzoni, 1980 and 1986; García, 1986). We described the brain as a complex neural oscillator composed of several coupled simpler oscillators. What fluctuates is the number of impulses fired by the neurons through the networks given by the histological structure. We focused on a relatively simple oscillator, which included excitatory and inhibitory endings in convergent, divergent and re-entrant pathways. The phenomenon has three scales arranged in increasing order: the time-scale of the after-discharge, the scale of the latency and the scale of the synaptic phenomena. The microscopic level in this case is in the time-scale of firing of each individual neuron. The afterdischarge emerges as the co-operative effect of the lower units (neurons), which being coupled together produce the scale of the oscillation. In the average description via the modulation equations for amplitude and phase, this scale emerges as the result of the interaction of an average mass and average stiffness. These properties account for the overall organization of the net.

The exchange of stability is attributed in our model to synaptic mechanisms. At this stage it is important to notice that the scale of the synaptic activity is so slow that it influences the macroscopic response. Higher units, such as cells, are described on the average, since their time-scale is microscopic compared to the ones of interest. The effects of the synaptic mechanism account for the modulation. The equations include a damping contribution which controls the shape of the after-discharge.

SOME UNSOLVED PROBLEMS

In the previous example, by using a detailed set of equations, it was possible to relate quantitatively the physiological parameters to the macroscopic features of the kindling effect developing in time. It remains to explain the spatial spreading of the effect. Some attempts in this direction will be discussed in this section.

The general problem is to study the motion of patterns of synchronization. In the case of the kindling effect, the idea is to explain the spreading of the effect to various secondary sites in the brain as a wave of synchronization which starts from the primary sites, and leads to a different site, imposing a synchronization structure along a pre-established path.

A possible solution is to model the spreading of synchronization as a *travelling wave* (García, 1986), such that in this case the 'phase' will account for the state of relative synaptic efficacy, and the problem can be formulated as one of propagation of a phase. Then, the propagation of these changes in phase can be described in terms of the modulation theory described in the Appendix. What is still lacking is the appropriate relation between local lengths and frequencies given by the dispersion relation.

Much remains to be done in this direction. In particular, the mechanisms of generation and stopping of the spreading patterns are not yet understood. In many ways this is a problem similar in nature to the generation of water waves and their breaking, whose mechanism is far from clear after hundreds of years of study.

Another unsolved problem concerning plasticity is how *spontaneity* arises in neural networks. By spontaneity is meant the capability of the nervous system to generate patterns of activity which are largely independent of perturbations occurring in the surroundings. There is another laboratory phenomenon related to the kindling effect which is useful to study the emergence of spontaneity in brain networks. This is the *dysrhythmic cerebral state* (Alonso deFlorida and Delgado, 1958; Alonso deFlorida and Minzoni, 1980). Spontaneous abnormal brain rhythms and self-sustained discharges (similar to the elicited afterdischarge) are the distinctive feature of this phenomenon. By using repeated stimulation trains (for one hour) instead of a single train as a stimulation procedure in the long scale, this phenomenon is produced instead of the kindling effect. This emergent oscillatory phenomenon remains for weeks or months and, unlike the kindling effect, it is associated with the emergence of conspicuously abnormal animal behaviour. The phenomenon resembles human temporal epilepsy. We are presently studying this phenomenon by using models previously elaborated to understand chaos in physics (Guckenheimer and Holmes, 1983). The hypothesis is that in such advances states of experimentally emergent neural organization, there are several coupled neural oscillators. Since the individual modules behave according to the Hopf bifurcation, one should expect the appearance of complex attractors of low dimensions when the modules interact. Such attractors can be characterized by solutions with no obvious order and no relation to external perturbations, which indeed could account for spontaneity. The main issue is to resolve experimentally the 'number of modules' involved in the process. The question of spatial organization, which is itself 'chaotic', remains to be resolved.

APPENDIX

We begin by giving a schematic description of how instabilities are handled using modulation theory. The basic idea is to take the oscillatory solution precisely at the point of exchange of stability and itnroduce a slow phase and amplitude. The slow phase and amplitude evolve according to average equations. The parameters of the averaged (modulational) equations are obtained via complicated nonlinear interactions of the parameters of the original system.

A model can be successful only if the parameters of the averaged equations can be related to the physiological parameters in the system. If this is the case, the averaged equations will be able to describe and predict macroscopic behaviour as arising from the lower level.

In this part of the appendix we give a schematic description of modulation theory. (For a more detailed presentation see Kevorkian and Cole, 1981). The aim is to use these ideas to describe the propagation of synchronisation patterns in the later stages of the kindling effect.

The elementary solutions of the equation

$$F(u) = 0$$

which model the system

Are periodic solutions $u = P(kx - \omega t)$ where P is a periodic function. The variable x is a space variable; k is the wave number fixing the space scale of the wave. The variable t is time and the frequency ω fixed the basic time scale of the system. The dynamics is embodied in the dispersion relation $\omega = \Omega(k)$ where the function depends on the system considered. It is shown that variations in k as a function of x and t which represent the modulation satisfy the equation of conservation of waves,

$$k_t + (\Omega(k))_x = 0$$

which describes the modulation due to initial conditions. Also, for non-homogeneous media, we have $\omega = \Omega(k, x)$ and the equation becomes:

$$k_t + \Omega'(k)k_x = -\Omega_x$$

where Ω_x is the effect of the inhomogeneities on the waves. The key information is the relation $\omega = \Omega(k)$ and this could be found without a detailed knowledge of the equation by global considerations such as isotropy, short and long-wavelength behaviour, etc. In the case of wave of synchronization, the proposed idea is to find a suitable Ω and try to correlate the wave patterns with the patterns observed in the spreading of the kindling effect.

REFERENCES

Alonso-deFlorida, F. and Delgado, J.M.R. (1958) *Amer. J. Physiol. 193,* 233.
Alonso-deFlorida, F. and Minzoni, A.A. (1980) In: *Limbic Epilepsy and the Dyscontrol Syndrome,* M. Girgis and J.G. Kilod (eds) p. 63. Amsterdam: Elsevier.
—— (1986) *J. Theor. Biol. 120,* 285.
Garcia, C. (1986) *Comunicaciones Tecnicas. Serie Naranja.,* No. 452. IMMAS, Universidad Nacional Autonoma de México.
Goddard, G.V. and Douglas, R.M. (1976) In: J.A. Wada, and R.T. Rodd (eds) *Kindling,* p. 1. New York: Raven Press.
Guckenheimer, J. and Holmes, A. (1983) In: *Applied Math. Sciences,* vol. 42; New York: Springer-Verlag.

Kevorkian, J. and Cole, J. (1981) *Perturbation methods in Applied Mathematics*. New York: Springer-Verlag.

Matthies, H. (1982) In: C. Ajmone-Marson and H. Matthies (eds) *Neuronal Plasticity and Memory Formation*, New York: Raven Press.

16. Formal languages and theoretical molecular biology

MIGUEL A. JIMÉNEZ-MONTAÑO

Thus we have the chicken–egg paradox in a new form: which came first, the language or the constraint?

Pattee (1972)

In his epilogue to the fourth volume of papers issuing from the IUBS Symposia at Villa Serbelloni, inspired by the papers of Pattee (1972) and Thom (1972) among others, Waddington concluded that:

The situations which arise when there is mutual interaction between the complexity-out-of-simplicity (self-assembly), and simplicity-out-of-complexity (self-organization) processes, are, I think, to be discussed most profoundly at the present time with the help of the analogy of language.

However, after fifteen years, very little has been accomplished to substantiate this statement.

Waddington conceived the Theory of General Biology as a discipline that is not so deeply interested in the understanding of any particular biological phenomenon for its own sake, but mainly in so far as it promotes a greater understanding of the general character of the processes that go on in living as contrasted with non-living systems. He suggested that it is a *language* that may become a paradigm of such a theory, but a language in which the basic sentences are programs not statements. To what extent have his expectations been fulfilled?

The answer most people would give is that the progress achieved is negligible; but, probably, few have asked themselves why? Perhaps one reason for this situation lies in the general feeling, expressed in the same epilogue, that none of those present at that symposium seemed to have felt much inclination to suggest that the crucial questions for a general theory of biology lie in the area of molecular biology.

However, it is precisely in this area, where the processes of self-assembly and self-organization are beginning to be understood, where the linguistic terminology has been universally adopted but, usually, at only a metaphorical level. As a result of the great advances achieved in the last years in the sequencing of proteins and nucleic acids, as well as in the determination of the three-dimensional structure of macromolecules by means of X-ray crystallography, there is a pressing need to rigorously establish the relationship between the 'information' contained in the sequence, and the corresponding spatial features which determine a given molecular folding pattern and the concomitant biological function. This problem of decoding the second half of the genetic code ('that for a long time people forgot it was a problem', Kolata, 1986) has recently attracted much attention. It is in such a context of characterizing the set of amino-acid sequences which are compatible with a given three-dimensional structure that the conceptual framework we are going to describe first arose.

As we shall see, there are still more questions than answers; but the possibility of posing the questions precisely and the experimental data available allow a confrontation between theory and experiment which, presently, is hardly possible in other areas of biology.

The theoretical as well as practical relevance of establishing the so-called stereochemical code, on which necessarily converge the informational and structural aspects of biological macromolecules, brings to the foreground the well known differences between the 'informationist' and 'structurist' schools of molecular biology (Stent, 1968) which, contrary to what one would think, are far from solved. Pattee (1979) points out that we now find that the tension between both schools of molecular biology, whose conceptual foundations had little in common, has been relaxed to a kind of loose acceptance of both structural and informational language that has very little to do with precise physical language in either case. Obviously, the way out of this situation will not simply be the result of the accumulation of more experimental facts. On the one hand, critical appraisal of our present ideas, which remain more or less hidden in the usual linguistic terminology, is in order. On the other hand, just to mention one example, the problem of establishing how different two structures can be and still be considered 'the same' has no clear objective answer at this time (Ponder and Richards, 1987).

Although the syntactic approach to the analysis of biosequences was introduced independently of Pattee's ideas (see Ebeling and Jiménez-Montaño, 1980; Ebeling and Feistel, 1982; and Jiménez-Montaño, 1984), its starting-point could well have been his statement (Pattee, 1972) that: 'The concepts of constraint and language are very general, and closely related at a deep level'. However, he does not explain how this idea should be developed in order to obtain a useful formulation. To make any progress in the establishment of Waddington's paradigm, it is first necessary to show (in any particular area of biology) that a proper abstraction of the concept of natural language produces a suitable theoretical framework for its description; as was done for computer languages after the work of Chomsky and others. Secondly, it is required that this formal language somehow encodes the physico-chemical constraints obeyed by the macromolecules under study. As shown with specific examples in Jiménez-Montaño and Martinez (1984) and Jimenéz-Montaño (1988) (see below), the syntax rules which characterize the sequences belonging to a protein, or RNA family, presumably encode the constraints which are mainly responsible for its common three-dimensional form. Furthermore, the advantage of having a formal theory is not only the possibility of making predictions but, also, to show its limitations.

STRUCTURAL LINGUISTICS

In his paper on structuralism and biology (Thom, 1972), Rene Thom asks himself if the structuralist developments in anthropological sciences (such as linguistics, ethnology, and so on) have a bearing on the methodology of biology? He answers this question affirmatively. According to him: 'its prospective application to biology lies between the old, Linnaean type of biology, descriptive and taxonomic, and the modern biology (molecular biology, physiology, ecology) of a more reductionist, explanatory nature'. Furthermore, he recommends that we should not underestimate the epistemological importance of such methods because, as he puts it:

> From a qualitative point of view, they represent an analogous position to the famous *Hypotheses non fingo* of Newton. In the structuralist viewpoint, one does not try to explain a morphology by reduction to elements borrowed from another theory, supposedly more elementary or fundamental, as one might try to explain biology by physics and/or chemistry, or sociology by psycology or biology; one only tries to improve the description of the empirical morphology by exhibiting its regularities, its hidden symmetries, by showing its internal unity through a formal mathematical model which can be generated axiomatically. In this respect 'structuralism' is a modest theory, as its only purpose is to improve description.

Formal language theory started around 1956 when Noam Chomsky gave a mathematical model of a grammar in connection with his study of natural languages. Although Chomsky's work represented a radical departure from previous linguistic theories, it arose within and was influenced by the structural approach developed by Zellig Harris, among others. It is a remarkable fact that the properties of language which Harris (1972) considered to be universal and essential, and which are relevant to a mathematical formulation are, apparently, also satisfied by biomolecular sequences (for example, the set of sequences belonging to a protein family such as the globins, say).

Without going into too much detail, I will now show how this knowledge from structural linguistics may help us to understand how the 'information' contained in a protein sequence codes for the corresponding three-dimensional structure and its associated biological function.

According to Pattee (1972), a molecule becomes a message only in the context of a larger system of physical constraints which he has called a 'language'. Leaving aside an abstract definition of language, which is independent of any communication concept, I consider here the idea of language in its everyday usage. (I will deal with abstract definitions in the next section.)

First, the elements are discrete and arbitrary. The only elements which are going to be relevant for the grammatical description, nucleotides or amino acid residues, are discrete ones. Their arbitrary character refers to the fact that, for example, the amino acids from which a short peptide (five residues or less) is composed do not suggest its biological function, not the conformation it will assume within a given protein. In this respect we may recall Thom's (1972) observation:

> In the 'genetics–linguistics' analogy, the Homologue of the nucleotide sequence of DNA is the phonemic structure of speech (or the alphabetic structure of writing). It is here, at this lowest level, that the Saussurian "arbitrariness of sign" is most conspicuous, here the effect of random fluctuations in the past history is most important. At the higher level of syntax, however, the word order is much less arbitrary, as it manifests a relatively tight connection with meaning.

Secondly, combinations of elements are linear, denumerable. Third, not all combinations of elements constitute a sentence. A precise definition of a sentence is given below in relation to a corresponding definition of formal grammar. Intuitively, we describe a 'sentence' (word-string) as a naturally occurring biosequence, assuming that it obeys the regularities for 'sentencehood' required by the 'genetic grammar'. Deleterious mutations may cause a sequence to deviate and become a semi-sentence (i.e. a sequence which is not completely grammatical).

In every natural or artificial language, not all finite sequences of elements (phonemes, nucleotides, amino acids) occur as sentences. As Harris (1972) explained for natural languages, the fact that not all combinations occur makes it possible to define larger elements (e.g. morphemes in linguistics and exons – i.e. segments of DNA which are converted to protein – secondary structure or super-secondary structure segments in DNA and proteins, respectively) as *restrictions* on the combinations or smaller elements. As happens in natural languages, so also in 'genetic' language is this *redundancy* essential, because the possibility of distinguishing the elements requires that not all combinations occur. For example, let us consider the set of peptides which adopt a definite spatial conformation (alpha-helix, beta strand) when immersed in a polypeptide chain (Wierenga et al., 1986), each of which has a different sequence of amino acids. If every sequence constitutes a secondary or super-secondary structural element, there would be no way of identifying such elements, or even of knowing where their boundaries lay within a protein (sentence), except by reference to a metalanguage or translation (i.e. to the experimentally determined three-dimensional structure of the protein). Like a natural language, the genetic language must contain considerable redundancy in its nucleotide and exon sequences.

If we call these larger elements in DNA and proteins, 'morphemes', we can see how the following argument of Harris (1972) extends naturally to the 'genetic language'. Harris writes:

> In all languages, the departures from complete utilization of morpheme sequences are such as to permit the setting up of morpheme classes. That is, it is always possible to collect morphemes into classes in such a way that the sentences or discourses of the language are far less redundant as a set of sequences of these classes than they are as a set of sequences of morphemes. The classes of morphemes are (or can be) defined in each language solely by this criterion.

Furthermore, he shows that in every language, not only is there redundancy in respect to the sequence of ultimate elements (phonemes, nucleotides, amino acids), but also this redundancy is composed of a system of contributory redundancies, each in terms of intermediate elements. Each of these contributions to the total redundancy has meaning: the meaning of entities, and the meaning of grammatical relations among them, is related to the restriction on combinations of these entities relative to other entities. Finally, he makes the following remark which, in agreement with what I said before, also applies to the genetic language 'this fits with the fact that the phonemes, which are not defined on the basis of redundancy of some other entities, do not have meaning'.

I have applied these ideas to the study of the genetic language with encouraging results (see below, and Jiménez-Montaño, 1984; Jiménez-Montaño, 1988). The concept of redundancy, which is defined in Ebeling and Jiménez-Montaño (1980) and in Ebeling and Feistel (1982), is a generalization of the usual Shannonian concept and refers not only to the frequency of occurrence of certain sequences, but to whether certain sequences occur at all as accepted sentences.

In concluding this section it can be said that the general problem of structural linguistics is the question of how to distinguish the sequences (of phonemes, morphemes, etc.) which occur as sentences of the language in contrast to sequences which do not occur as sentences. Following Harris (1972), it could also be said that it is a problem of finding regularities in those sequences of elements which constitute sentences as contrasted with those which do not. This is the problem that has to be solved in order to understand how sequence 'information' codes for structures which carry biological functions. It should be noted, however, that for all languages, including the genetic language, any attempt to distinguish the word sequences which are sentences from those which are not has to satisfy the condition that the boundary between the two is not sharp. In the latter case this is due to the problem of determining the 'sameness' of two structures, as I mentioned at the beginning of this paper.

Formal languages and formal grammars

In so far as natural languages have certain properties (duality of structure, recursivity, grammatical ambiguity, etc.) which they share with particular kinds of formal languages, some linguists consider that formal languages may usefully serve as models of natural languages (Lyons, 1978). The same assumption is made here with respect to the genetic language. It is important to realize, however, that the notion of a formal language is totally independent, in principle, of the notion of a natural language. It is also a much more general notion. The syntax of a formal language can be described without reference to any interpretation that might be assigned to the elements or combinations of elements. As was clearly emphasized by Lyons (1978):

> A formal language might serve as a model in principle, for all sorts of systems that have nothing to do with communication and would never be described as languages in the everyday sense of the term.

It is beyond the scope of this paper to give even the bare rudiments of the theory of formal languages. Therefore, we refer the interested reader to the excellent accounts found in the literature. A brief and very readable summary may be found in Appendix 1 of Lyons' book (1978); another short reference is Spanier (1969). More complete presentations may be found in Salomaa (1973) and Hopcroft and Ullman (1969). However, to facilitate the understanding of the results reported in the next section, we include here some basic definitions.

A formal language is defined abstractly as a mathematical system. An *alphabet* or *vocabulary* is any finite set of symbols. Some examples of alphabets are the Latin alphabet (A,B,C, . . . ,Z), the DNA alphabet (A,C,G,T), the RNA alphabet (A,C,G,U), and the protein alphabet (A,C,D, . . . ,Y,W). In the latter the letters represent the twenty amino acids in the standard single-letter notation. A *sentence* over an alphabet is any string of finite length composed of symbols from the alphabet. The empty sentence, E, is the sentence consisting of no symbols. If V is an alphabet, then $V^\star$ denotes the set of all sentences composed of symbols of V, including the empty sentence. Thus, if $V = (0, 1)$, then $V^\star = (E, 0, 1, 00, 01, 10, 11, 000, . . .)$. A *language* is any set of sentences over an alphabet. One way to represent a language is to give an algorithm which determines if a sentence is in the language or not. Such an algorithm is said to *recognize* the language. Here we are more interested in representing languages from a *generative point of view*. That is, we need a procedure which systematically generates successive sentences of the language is some order. Such a procedure is called a *grammar*. A grammar has two tasks: it must enumerate each sentence; and not be a non-sentence. It must also associate with each sentence a grammatical analysis which can provide the kind of structural information (in the sense of structural linguistics) needed to explain examples. This structural description refers to the elements the sentence contains, their relations to each other, the relations of the sentence to other sentences, and so on (see Postal, 1964; Chomsky, 1964). For phrase-structure grammars (see below) it corresponds to the familiar idea of diagramming or parsing an English sentence. Formally, we denote a grammar G by (V_N, V_r, P, S). V_N is the set of syntactic categories, which in natural languages are 'singular noun', 'verb phrase', etc. They are called 'non-terminals' or 'variables'. V_T is the set of objects which play the role of words or letters of the alphabet, according to the level in which the analysis is conducted. P refers to the relation that exists between various strings of variables and terminals. These relationships are called 'production rules'. For example, $\langle$noun phrase$\rangle \rightarrow \langle$adjective$\rangle \langle$noun phrase$\rangle$. Finally, one non-terminal is distinguished, in that it generates exactly those strings of terminals that are deemed in the language. It is called 'start' symbol, S (see Hopcroft and Ullman, 1969). Different grammars are defined by the form of the production rules. The grammars we have applied to the study of the biosequences belong to the class known as phrase-structure grammars. Such grammars consist of a finite set of rules of the form: $X A Y \rightarrow X Z Y$, where the arrow is to be interpreted as the instruction 'is to be rewritten as', and the following conditions apply:

Condition 1) X, Z, and Y are strings of symbols (X or Y or both possible void) but A is a single symbol;

Condition 2) Z is not void;

Condition 3) A is not identical with Z.

In the above rules X, Y represent the context for the expansion. If we require that X and Y are void, then if a symbol A is expanded into a string Z it is expanded into Z in every sequence in which it occurs. Such grammars are called context-free. Although it is almost certain that the grammars needed for describing protein families are context-sensitive, the applications of formal grammars to biosequences that we wish to describe are in terms of context-free grammars only. Postal (1964) noticed that conditions 1–3 constitute the only precise method known for assigning structural descriptions to an infinite set of sentences. They were originally proposed by Chomsky and although they are necessary they are not sufficient. Postal added a fourth condition:

4) Permutations are excluded.

This means that the notion of phrase-structure grammar implies that pairs of rules of the form AB $\rightarrow$ BB and BB $\rightarrow$ BA are forbidden.

In this respect it is interesting to note that by the process of 'alternative splicing' isoenzymes are found to differ in the exonic regions from which they are composed. (For a description of the mechanism of alternative splicing – for the generation of multiple protein isoforms from single genes – see Breitbart *et al.*, 1987.) However, no exon permutations have been observed. This is understandable because such permutations of substrings would produce radical changes in tertiary structure of the proteins.

With the help of the above definitions, we can move on to formulate the problem of interest in precise terms. This is the problem of inferring a grammar for a given set of strings (Cook *et al.*, 1976). In his original formulation, Chomsky (1957) put the goal of grammatical inference as obtaining a *simple grammar* whose language fits the given sample. For our purposes the goal is somewhat different. We are not only interested in fitting the given sample, but in fitting the class of sequences belonging to a protein or RNA family (i.e. which correspond to a given fold) within pre-established limits. For us the object of the grammar is not simply to find a shorthand for describing the sample set, but to extend it in order to predict new sequences from the same family.

APPLICATIONS

I have applied the syntactic approach to the characterisation of sequences belonging to the globin superfamily (alpha-haemoglobin, beta-haemoglobin and myoglobin), to immunoglobin domains and t-RNA molecules.

I describe the protein applications first. In order to introduce physico-chemical constraints into a grammar, I have proposed a hierarchy of amino-acid categorizations (Jiménez-Montaño, 1984). Other sets of amino acid groupings have been proposed by Sneath (1966) and Lim (1974). These hierarchies can be represented by inverted trees, with the starting symbol or amino-acid category at the root. Both in Lim's and in my hierarchy this category is divided into two groups: hydrophobic amino acids (1) and non-hydrophobic (0). In this way a (two-symbol) reduced alphabet is obtained. Further splitting of the non-hydrophobics produced the neutral and the hydrophilic groups. Taking into account (besides hydrophobicity) the size of the residues, a six-group categorisation is obtained in my hierarchy. Continuing with this process we end up with single amino acids in the leaves of the tree. Other reduced alphabets in terms of chemical, functional, charge and structural properties have been studied by Karlin and Ghandour (1985); also by Dayhoff (1978), Miyata *et al.* (1979) and Taylor (1986), among many others.

Having selected the syntactic categories for the description, one infers a grammar for the protein family under study, as follows:

First, take a sample of sequences from the family and translate it into the reduced alphabet to be used (e.g. the binary alphabet). After aligning the translated sequences by means of standard alignment procedures, construct a context-free grammar by means of an algorithm (called 'grammar') that defines the re-writing rules in terms of the most frequent occurring pairs, triplets, quadruplets, etc. in the (translated) sample. In this way a non-redundant description of the sequences, in the class to which the sample belongs, is obtained. With each sequence the grammar associates a parsing, in terms of these sub-sequences.

In the case of the globins we have been able to separate them into different clusters, corresponding to the different families, each one with its characteristic sub-sequences of zeros and ones. If a more detailed description is required, the procedure is repeated, for example, using a six-letter alphabet. If the description were attempted in the full twenty-letter alphabet, no regularities would be found, since two globins may have a degree of homology as low as 16 per cent. Similar results of clustering were also obtained for the immunoglobulin domains.

Obviously, there are many ways of splitting the amino acids in two groups. How is it possible to know if one reduced alphabet is better than another? Jiménez-Montaño, Zamora-Cortina and Trejo-Lopez (1988) proposed an optimization principle to compare amino-acid categorizations. The best categorization with respect to a given sample is the one that under certain conditions removes the most redundancy.

Multiple alphabets have also been employed for the analysis of DNA sequences. Karlin and Ghandour (1985b) proposed three two-letter nucleitide alphabets: S–W alphabet W = A or T; S = G or C), where W stands for weak bonding of bases and S for strong bonding of bases; P–Q alphabet which distinguishes purines (P = A or G) from pyrimidines (Q = C or T) and a 'control' E–F alphabet (E = A or C; F = G or T) which does not appear to have any natural chemical association.

With T replaced by U, I have used these nucleotide categorizations for the analysis of tRNA sequences (Jiménez-Montaño, 1988). By means of the program 'grammar', which as explained above finds the most frequently occurring subsequences, I have found what appears to be a primordial tRNA sequence. It is constructed using as 'building blocks' the self-complementary word CCGG, the word AACC and its complement GGUU, and scattered pairs and single nucleotide substitutions which make it as similar as possible to the 'master' sequence, obtained from the alignment of 144 tRNA sequences. The only major deviation from this 'master' sequence occurs in the triplets before and after the anticodon.

I have also built a 'general tRNA grammar' which takes into account, besides the above categorizations, the structural information encoded in the clover-leaf secondary structure and twenty-three 'constant' positions (plus anticodon) involved in tertiary interactions, as observed in yeast tRNA (phe). By applying the rules of this grammar, one may parse only those strings in the corresponding tRNA language and no other string. These strings encode the secondary and tertiary constraints. Therefore, in this case, the 'structural description' (in the sense of structural linguistics) nicely coincides with what the biochemist understands for this term.

Since most tRNA molecules have seventy-six bases, the space a priori associated with these molecules consists of $4^{76} \approx 5.7 \times 10^{45}$ possible strings. It is reduced by the grammar to a language consisting of, approximately 7×10^{19} syntactically correct sentences. For further results I refer the reader to the original paper (Jiménez-Montaño, 1988).

FINAL REMARKS

Besides the models described in this work, other authors have also applied formal language theory to the study of biological macromolecules. Brendel and Busse (1984) have expressed the process of the translation of messenger RNA into proteins by means of finite-state transducers and have derived conclusions using the closure properties of the class of regular languages; algorithms for string comparisons have been analysed and applied to the handling of sequences data for macromolecules in an extensive literature (for a review see Martinez, 1984). Head (1987) has applied formal-language theory to DNA by making an analysis of the generative capacity of specific recombinant behaviours. All these developments constitute only the first efforts towards the understanding of the 'genetic language'. Crude and incomplete as they are, they nevertheless represent new directions towards a theoretical biology.

I have hinted here at some of the complicated ways in which the concepts of 'constraint' and 'language' are deeply interwoven. The constraints are necessary for the molecules to fold in a proper way; but biological macromolecules are the result of natural selection acting on a global system that Pattee has called a 'language'. Which came first, the language or the constraint? The problem, I think, still has to be resolved.

REFERENCES

Breitbart, R.E., Andreadis, A. and Nadal-Ginard, B. (1987) Alternative splicing: a ubiquitous mechanism for the generation of multiple protein isoforms from single genes. *Ann. Rev. Biochem. 56*, 467–95

Brendel, V. and Busse, H.G. (1984) Genome structure described by formal languages. *Nucleic Acid Res. 12*, 2561–8.

Chomsky, N. (1957) *Syntactic Structures*. The Hague: Mouton & Co.

Chomsky, N. (1964) On the notion 'Rule of Grammar'. In J.A. Fodor and J.J. Kats (eds) *The Language* pp. 119–136. Englewood Cliffs, N.J.: Prentice-Hall Inc.

Cook, G.M., Rosenfeld, A. and Aronson, A.R. (1976) Grammatical inference by hill climbing. *Informational Sciences 10*, 59–80.

Dayhoff, M.O. (1978) *Atlas or Protein Sequence and Structure*, Vol. 5, suppl. 3, Washington, D.C.: National Biomedical Research Foundation.

Ebeling, V.W. and Feistel, R. (1982) *Physik der selbstorganisation und evolution*. Berlin: Akademie-Verlag.

Ebeling, V.W. and Feistel, R. (1982) *Physik der selbstorganisation und evolution.* Berlin: Akademie-Verlag.

Ebeling, W. and Jiménez-Montaño, M.A. (1980) On grammars, complexity and information measures of biological macromolecules. *Math. Biosci. 52*, 53–71.

Harris, Z. (1972) *Mathematical structures of language*, pp. 6–19. Interscience tracts in pure and applied mathematics, number 21. London, Sydney, Toronto: John Wiley & Sons.

Head, T. (1987). Formal language theory and DNA: an analysis of the generative capacity of specific recombinant behaviours. *Bull. Math. Biol. 49* (), 737–59.

Hopcroft, J.E. and Ullman, J.D. (1969) *Formal Languages and their Relation to Automata.* Reading, Mass.: Addison-Wesley.

Jiménez-Montaño, M.A. (1984) On the syntactic structure of protein sequences and the concept of grammar complexity. *Bull. Math. Biol. 46*, (4), 641–59.

Jiménez-Montaño, M.A. (1988). On the syntax analysis of transfer RNA and a model for a primordial molecule. *J. Mol. Evol.* (submitted for publication).

Jiménez-Montaño, M.A., Zamora-Cortina, L. and Trejo-Lopez, J. (1988) Principio de optimizacion para categorizar elementos primitivos en el reconocimiento sintactico de patrones. *Rev. Aportaciones Matematicas, SMM.* Mexico D.F. (in press).

Jiménez-Montaño, M.A. and Martinez, H.M. (1984) A procedure for characterizing the primary structure of protein family. *J. Mol. Evol.* (unpublished manuscript). A new version of this paper is in preparation.

Karlin, S. and Ghandour, G. (1985a) Multiple-alphabet amino acid sequence comparisons of the immunoglobulin k-chain constant domain. *Proc. Natl. Acad. Sci. 82*, 8597–601.

Karlin, S. and Ghandour, G. (1985b) The use of multiple alphabets in kappagene immunoglobulin DNA sequence comparisons. *EMBO J. 4*, 1217–23.

Kolata, G. (1986) Trying to crack the second half of the genetic code. *Science 233*, 1037–9.

Lim, V.I. (1974) Algorithms for prediction of alpha-helical and beta-structural regions in globular proteins. *J. Mol. Biol. 88*, 873–94.

Lyons, J. (1978) *Chomsky* (Fontana Modern Masters) Glasgow, UK: William Collins & Sons.

Martinez, H.M. (ed.) (1984) Mathematical and computational problems in the analysis of molecular sequences. *Bull. Math. Biol.* (special issue honouring M.O. Dayhoff), *46*.

Miyata, T., Miyazawa, S. and Yasunaga, T. (1979) Two types of amino acid substitution in protein evolution. *J. Mol. Evol. 12*, 219–36.

Pattee, H.H. (1972) Laws and constraints, symbols and languages. In: C.H. Waddington (ed.) *Towards a Theoretical Biology 4. Essays* pp 248–258. Edinburgh: Edinburgh University Press.

Pattee, H.H. (1979) The complementarity principle and the origin of macromolecular information. *Biosystems 11* (2,3), 217–26.

Ponder, J.W. and Richards, F.M. (1987) Tertiary templates for proteins. Use of packing criteria in the enumeration of allowed sequences for different structural classes. *J. Mol. Biol. 193*, 775–91.

Postal, P.M. (1964) Limitations of phase structure grammars. In J.A. Fodor and J.J. Katz (eds) *The Structure of Language.* Readings in the Philosophy of Language, pp. 137–51. Englewood Cliffs, N.J.: Prentice-Hall.

Salomaa, A. (1973) *Formal Languages.* New York: Academic Press.

Sneath, P.H.A. (1966) Relation between chemical structure and biological activity in peptides. *J. Theor. Biol. 12*, 157–95.

Spanier, E. (1969) Grammars and languages. *Amer. Math. Monthly 76*, 335–42.

Stent, G. (1968) That was the molecular biology that was. *Science 160*, 390–5.

Taylor, R.W. (1986) The classification of amino acid conservation. *J. Theor. Biol. 119,* 205–18.

Thom, R. (1972) Structuralism and biology. In C.H. Waddington (ed.) *Towards a Theoretical Biology 4. Essays* pp. 68–82. Edinburgh: Edinburgh University Press.

Wierenga, R.K., Terpstra, P. and Hol W.G.J. (1986). Prediction of the occurrence of the ADP-binding $\beta\alpha\beta$-fold in proteins, using an amino acid sequence fingerprint. *J. Mol. Biol. 187,* 101–7.

17. Towards a grammatical paradigm for the study of the regulation of gene expression

JULIO COLLADO-VIDES

This paper is about the various attempts pursued in the advancement of biological knowledge from a theoretical standpoint. Most formalization attempts in biology are based on the application of concepts or methods taken from the fields of physical chemistry or mathematics. The challenge for those of us who believe in the possibility of a theoretical biology, is to achieve the right combination between the contribution of those fields of science which are more structured than biology and the complex processes with which biology is concerned, such that the biological sciences will be ultimately enriched. One of the risks regarding these attempts, is portrayed in the opinion which states that the theoretical scientist spends his time 'playing mathematical games'.

A science without theory is non-existent. Although in a lesser degree than in physics, biology has its basic concepts and even complete theories such as those concerning evolution (natural selection, neoteny, etc., Gould, 1977). These types of concept and the ideas behind them provide a sample of the theoretical developments in biology.

Nevertheless, the advantage currently enjoyed by the accumulation of experimental information over its theoretical counterpart is quite evident. Some important reasons for the difficulty in creating formal theories in biology are: (a) the great diversity found in the biological world, which easily generates exceptions to any general explanation; and (b) the complex nature of biological processes.

OBJECTIVE

The approach used in this paper, in contrast to the usual physics-oriented approach used in theoretical biology, draws its inspiration from the field of generative grammar which studies a highly diverse and complex object: human language.

In effect, human language shares the above mentioned characteristics with molecular biology. The diversity of human tongues is enormous and the diversity within one particular tongue is likewise gigantic. This fact provides for the appearance of 'exceptions' to almost any proposed law. In like manner, human language is a complex system (van Riemsdijk and Williams, 1986). Perhaps it is this coincidence which explains the fact that generative grammar, whose methodology has mathematical roots, has also been accused of playing 'mathematical games'.

The basic purpose of this paper is to present the fundamental ideas regarding the generative grammar treatment of genetic data at a molecular level, such that none of the aspects presented here will be analyzed in great detail. The justification and usefulness of the grammatical paradigm's various basic points, however, will be clearly outlined. We shall illustrate linguistic concepts in the description of prokaryotic regulation units.

GENERATIVE GRAMMAR CHARACTERISTICS

For our purposes in this project, we shall list four important characteristics of generative grammar:

1. Generative grammar is a discipline which searches for the laws that govern the combination of a finite group of symbols to generate an infinite group of sequences. It is a theory which tries to explain not a 'corpus' of data, but rather an eminently creative capacity: that of the human language. This constitutes one of the most valuable properties in the search for a theoretical development of molecular biology, which ultimately depends on processes created by evolution.

2. Because of its theoretical nature, generative grammar concepts are far from describing the 'physical' character of language as a human activity. Generative grammar is concerned with the study of the human capacity for speech, rather than the physical aspects involved in its expression. It therefore studies an abstract property specific to the human brain. As such, it has been compared to physics, in the sense that its theory and data are clearly separated.

3. Although its data and concepts are clearly separated, generative grammar studies human language by generating hypotheses which are tested 'experimentally' and which may be modified by syntactic arguments. This makes the approach offered by generative grammar to be a theoretical–experimental methodology.

4. Generative grammar is the linguistic approach which most emphasizes the pursuit of the biological (genetic) component of human language (Lightfoot, 1982); furthermore, as previously stated, it uses a methodology born from the mathematical treatment of languages.

To avoid repetition, we shall state the linguistic hypotheses directly in terms of molecular biology, thus explaining the concepts from a biological perspective. For a formal development of some of the ideas presented here, see Collado-Vides (in press). I will present the basics underlying a grammatical paradigm for the regulation of gene expression, which may be applied to a great diversity of genetic data at a molecular level.

WHAT IS A GRAMMAR?

A theory is a group of principles which can be used: first, to describe experimental data adequately; second, to generate explanations for the data; and, third, to generate predictions concerning unknown aspects of the data. The basic theoretical tool for the study of language is grammar.

A grammar is a theory composed of a group of rules and a group of principles which govern the use of these rules. The linguistic generative theory also requires other qualifications (Chomsky, 1975):

1. A *membership criterion* which defines which sentences are part of the language and which ones are not; that is, a criterion for data validation which must be external to the grammars themselves.

2. An external *comparison criterion* between the possible grammars (i.e. one grammar is preferable to another, inasmuch as it provides a better fit to the data). The best fit will be that which generates *all* of the language's chains *and only those* chains; these conditions serve as two 'adjustment knobs', one for the superior limit and one for the inferior limit.

Furthermore, there exists an internal criterion for grammar selection based on the simplicity, elegance and various arguments that have been elaborated for the different components (types of rules) of human language (Jackendoff, 1977; Emonds, 1976).

In the study of human language, there are two basic approaches depending on the membership criterion which is chosen. Those who study *what man speaks* are said to study the productive or acting aspects of language. Those who study *the human capacity of speech* are said to study the competence aspects of human language. Generative grammar is a theory of the competence of human language.

HOW IS A GENERATIVE GRAMMAR BUILT?

A grammar is similar to a family of equations obtained from a group of data or points in space. When we manage to formulate the equation or analytical expression, we obtain a predictive and explanatory capacity over certain aspects of the data.

A grammar is built (and rebuilt when necessary) by the combination of two stages. The first is the inductive stage, in which the data are assigned a particular grammatical description. The particular descriptions of different sentences (experimental points) are compared to obtain common rules for several sentences, until a reduced description of the group of data is obtained. A grammar to be used as a working hypothesis has thus been built and the deductive stage now begins, in which the grammar generates the original data and generates new expressions not included in the data. These predictions must then be tested experimentally to begin a new theoretical–experimental cycle of enrichment.

The great advantage of such a theoretical construction lies in the fact that the 'bottom-up' construction of the grammar will instil the data's structure into the grammar's form and, therefore, it becomes difficult to fall into 'mathematical games' lacking in experimental significance. The disadvantage is that extremely '*ad hoc*' descriptions of the data may be elaborated, thus sacrificing the pursuit of generality.

The points are the sentences

For the linguist, the points in space in the above mentioned analogy correspond to the different sentences recognized by the native speaker (i.e. one who speaks the given tongue as a first language). In the theoretical approach for the study of molecular biology we have proposed (Collado-Vides, in press), the data are the different arrangements that may occur in the genome organization of any organism. More specifically, the data are the various genome 'transcription or regulation units'.

As we shall see, grammar associates a derivation tree to each transcription unit or 'genetic sentence'. Such a tree or syntactic structure is defined by the designation of the respective syntactic categories to the words of the sentence, as well as by the establishment of rules which determine the left-to-right order of occurrence of the words.

The equation is the grammar

Grammar as a theoretical device is formed by grammatical rules which operate over symbols, thus generating new symbols. The derivation or *generation* of a sentence in generative grammar begins with the use of an initial symbol S – as in sentence – which is rewritten according to the application of the first rule over, for example, a pair of symbols NP and VP (nominal and verbal phrase). Such symbols (or syntactic categories) are then likewise rewritten as other new symbols and so on, until the syntactic categories (or non-terminal symbols) are rewritten as 'lexical categories', i.e. symbols which are substituted by words of the sentence (terminal symbols). Some lexical categories in genetics are, for example, promoter, operator, structural gene, etc.

The ordered sequence of a derivation can be graphically represented by a derivation tree. A single derivation tree may generate many different sentences, just as an equation generates many specific solution points. For example, the substitution of the Pr (promoter) symbol, in a given tree, by 'TTAGCGATCCTACCTGACGCTTTT TATCGCAACTCTCTAC- TGTTTCT' or by 'TAGGCACCCCAGGCTTACACTTTA TGCT- TCCGGCTCGTATGTTGTG', or else by 'TAACACCGTGCGTGT- TGACTATTTT ACCTCTGGCGGTGATAATGGT', the ara BAD *E. coli:*, the lac *E. coli*, and the lambda phage promoters (Hawley and McClure, 1983), respectively, will generate different molecular sentences; just as the substitution of the 'Name' category by 'horse', 'boy' or 'table' will generate different English sentences. A syntactic structure assigned to a sentence is basically a derivation tree.

THE MEMBERSHIP CRITERION IN MOLECULAR BIOLOGY

As stated before, linguistic theory requires a criterion to decide if the data belong to the language, a criterion which should be independent of the grammars elaborated during the study of the language.

Which is the membership criterion of genetic language, corresponding to the grammaticality defined by the native hearer's–speaker's linguistic intuition in human language?

It is important to note that such a criterion will, in a fundamental manner, limit and characterize the extent to which linguistic theory may be applied to biology. We are searching for a criterion which will enable us to find laws which are independent of any particular organism, organ tissue, cell type or chemical reaction. Likewise, we search for laws, rules and restrictions of a clearly biological nature, distinct from the chemical and physical restrictions that every biological system must satisfy.

Within the study of the competence of human language, theories do not attempt to explain the limitations or the fragments of sentences produced by slips of the tongue, memory errors, etc. A similar abstraction in the development of the grammatical paradigm in molecular biology should help to highlight the fundamentally biological characteristics of genetic language, separating the criterion of *molecular grammaticality* from those restrictions that arise from strictly physico-chemical or chemical considerations.

On the other hand, the membership criterion must be independent of the grammar and *operationally* definable. Let us see if something of the sort actually exists.

Consider the following chains:

$$\text{Pr, Op, Ic, S, Tc} \tag{17.1}$$

$$\text{Op, S, Pr, Ic, Tc} \tag{17.2}$$

$$\text{S. Op, Ic, Pr, Tc} \tag{17.3}$$

where Pr = promoter, Op = operator, S = structural gene Ic = initiation codon, Tc = termination codon. The order of the sequences is presented such that RNA polymerase would read from left to right.

We can proceed to state without difficulties, making use of basic molecular biology concepts, that unlike the orders of chains (17.2) and (17.3), the order of chain (17.1) is viable and must belong to the language.

These simple cases clearly illustrate that genetic language has at least one criterion for defining what belongs and what doesn't in the language. There will be other cases in which the membership of the chain will not be so easy to determine and perhaps even the designing of an experiment will be, therefore, necessary.

It is important to note that, as in human language, the membership criterion (that of competence, not of performance), does not limit the language to pre-existing chains, be it written or spoken in the case of human language; or likewise, in the biological case, within the genome of a certain organism. We may extend the membership criterion to new sequences, which have not been generated yet and, in fact, to sequences of infinite length which, from a theoretical standpoint, it would seem reasonable to consider as part of the biological system. For example, consider the following infinite sequence case:

$$\text{Pr, Op, (Si)n} \hspace{3cm} (17.4)$$

where (n) means that the elements between parentheses may be repeated an indeterminate (not zero) number of times. Leaving aside chemical and physical restrictions – which would decrease the probabilities of transcription as S increases – such an *infinite* sequence may theoretically be considered as belonging to the biological world.

What is the criterion used to define membership? The criterion which ultimately defines membership in the biological world is natural selection, but this is not easy to define operationally.

The operational criterion I use to determine the acceptability of chains (17.1) and (17.4) and the non-acceptability of (17.2) and (17.3) (that is, that which defines molecular 'grammaticality') I consider to be chain 'regulability'. In effect, I am selecting 'regulation units'. More specifically, it may be thought of as an 'expressibility' criterion which selects 'transcription units'.

We propose 'regulability' as the molecular grammaticality criterion. In effect, regulation is a central concept of biological mechanisms which extends through all hierarchies of biological organization, from ecosystems to molecules; therefore, it is a criterion which satisfies the previously stated requirements of independence from any particular biological structure (organism, cell, molecular, etc.). Furthermore, it is an operationally definable criterion, given the possibility it offers of designing experiments to determine, for example, whether a given operon responds to a specific signalling metabolite. At the same time, this will initially limit the linguistic analysis to those regions of the genome which participate, in some way, in the formation of 'transcription units'.

Regulability is ultimately based on the chemical and biochemical principles which underlie all biological processes, just as 'the language' studied by generative grammar is actually an idealized part of every day language.

REGULABILITY IS A SYNTACTIC CRITERION

Note that the regulability criterion works at a level constituted by combinations of what in grammar are known as lexical categories (Pr, Op, etc.), in a manner independent of the specific 'words' to which they may correspond. If, instead of description (17.1), we used this same sentence rewritten as a sequence of nucleotide bases, it would not be as easy to determine its membership in genetic language. The same thing happens in any spoken language, where the grammaticality is intuitively perceived by the speaker, through the use of categories such as Name, Verb, Nominal Phrase, Verbal Phrase, etc. However, the speaker would be unable to perceive this grammaticality at a level of individual letters if he were ignorant of their syntactic information. It is relatively easy to realize, therefore, that in genetic language at a nucleotide base level, no operational membership criterion exists; a syntactic study at such level seems therefore improbable.

On the other hand, we do not need to specify which is the actual structural gene in (1) to determine its regulability. Regulability and physiological usefulness are not synonymous. For example, consider a fragment of the *lac* operon, chain (1) rewritten as

$$\text{Pr}lac, \text{Op}lac, \text{Sz} \qquad (17.5)$$

where the *lac* sub-index specifies the lac operon promoter and operator, and z specifies the structural gener for galactosidase. In this case, we have a 'genetic sentence' with physiological usefulness, *in addition* to being subject to regulation. However, using genetic engineering tools we could substitute the Sz gene for one from the *trp* (tryptophan) operon. For example, we could substitute the *trpE* gene which encodes for the I component of the anthranilate synthase enzyme, such that:

$$\text{Pr}lac, \text{Op}lac, \text{S}trpE \qquad (17.6)$$

This sentence would be subject to regulation, but would probably lack physiological usefulness, due to the fact that such bacteria would induce the synthesis of enzymes concerned with tryptophan metabolism in response to the presence of lactose within the cell. For a sentence to have physiologically useful information, regulability is a prerequisite; not so the other way around – we may have 'sentences' subject to regulation but lacking physiological meaning.

A similar phenomenon occurs in human language. We have syntactic or grammatical sentences which, nevertheless, may not satisfy the criteria of correct semantic formation. The famous 'colorless green ideas sleep furiously' only satisfies syntactic criteria (Chomsky, 1957), whereas, 'green furiously ideas sleep colorless' doesn't satisfy either of the two criteria.

In the general model presented (Collado-Vides, in press), the evaluation of physiological usefulness at a molecular level requires an additional grammar component, corresponding to the semantic component in the study of natural language.

One of the advantages of a membership criterion based on syntax is that it enables us to construct a prediction generating theory useful for the genetic engineer, who can discover profitable combinations for mankind, which may represent aberrant conditions in nature. (Consider the production of human hormones by a gene inserted into bacteria.)

In the following section, we shall discuss two basic relations present in every transcriptional unit and which correspond, from a linguistic standpoint, to rules of phrase structure.

Dominance and precedence relations in genomic organization

The grammar we have used for the study of gene expression regulation has two basic components, each of them composed of a specific group of grammatical rules. All rules concerning phrase-structure are applied first, followed by the application of the transformational rules.

We shall now see that certain characteristics of genetic information organization are naturally described by phrase-structure rules.

The form of a phrase-structure rule is

$$X \rightarrow Y \tag{17.7}$$

which is read as 'rewrite X as Y'. A group of such rules applied one after the other generates a 'derivation tree'. For example, for the *lac* operon we could propose the following tree:

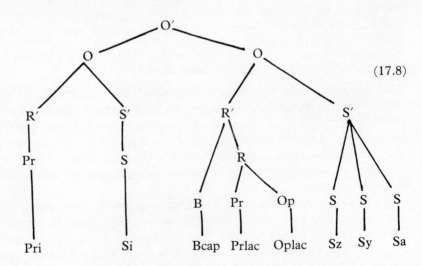

$$(17.8)$$

where the lexical and syntactic categories are: B = binding region; Pr = promoter; Op = operator; S = structural gene; R = regulation region (within an O); O = operon.

The specific items of the *lac* operon are labelled according to the following subindexes: i = repressor gene; cap = activation protein. These symbols are abbreviations for fragments of DNA sequences; such that 'Prlac', for example, corresponds to the 'TAGGCA . . . GTG' chain previously mentioned.

The prime notation is used in 'the theory of phrase-structure rules' (Jackendoff, 1977), to indicate that X′ is a projection of X. For example, a nominal phrase (N′) must have a name (N), and a verbal phrase (V′) must always have a verb (V). The obligatory character of the main category is one of the conditions in the elaboration of a derivation tree which comes from this theory.

This type of rule establishes two basic linguistic relations. The first one is the *dominance* relation which establishes that every word must belong to a lexical category and that certain arrangements of words must be constituents (syntactic categories) of some type. In (17.8) we have, for example, that 'Oplac' *is* an operator, 'Bcap,Prlac, Oplac,Ez,Ey,Ea' *is* an operon, etc. We consider that there is no doubt about the existence of dominance relations in the organization of transcription units, as shown by the existence of different lexical categories (promoter, operator, structural gene, attenuation sequences, termination signals, etc.).

The second relation is that of *precedence* which establishes a preferable order in the relative localization of words from left to right. Some precedence relations are, for example: i) 'Pr' is always present at the beginning of transcription units, ii) Translation initiation and termination signals are always within a structural gene.

The existence of precedence and dominance relations within genome organization shows how the molecular biologist uses notions around which linguists have elaborated an entire theory. Although there is a considerable distance separating human and genetic languages, these conceptual coincidences clearly portray the potential usefulness of generative grammar as a theoretical tool for the study of molecular genetics.

The proposed criterion which defines membership in genetic language is regulability. In the next section we shall see how a grammatical representation of regulation loops has been established.

GENERALIZATION OF THE STRUCTURE–FUNCTION RELATION

The structure–function relation is fundamental to all experimental sciences. In molecular biology the importance of the structure–function relation is clearly illustrated in the search for the determination of the 3-D structure of proteins (which ultimately determines their chemical function) based on their linear amino acid sequence. The linear sequence is itself determined by the corresponding genomic sequence of nucleotide bases.

By generalizing this type of relations, we can suppose that the organization of transcription units in the genome determines in an important way the regulation of the information contained in these units, given the fact that all DNA-interacting proteins are encoded by the DNA itself.

In the previous section, we saw how phrase–structure rules enable us to adequately describe information contained within the organization of the genome. We mentioned that the grammar we have used has two components, each formed by a particular group of grammatical rules. The first component is made up of phrase–structure rules, and the second, which we shall now illustrate, by transformational rules.

The transformational component used in the molecular linguistic approach is based on the hypothesis which states that the structure of a genetic sentence will determine the loops established for its regulation.

This hypothesis, in a similar manner to the theoretical elaboration of Universal Grammar in the study of human language, stems from a theoretical representation derived from the use of phrase–structure rules. In order for the transformational component to be implicitly determined by the organization of the genome, representations similar to (17.8) must be generated with some additional L sites or categories, which are empty (without words). Transformational rules are only able to displace words from their original positions to these empty spaces.

A genetic 'sentence' will have a finite number of L (as in loop) categories indexed as pairs, in which one of each pair will be empty. Thus, for example, the regulation of DNA by an allosteric repressor protein P (see Figure 17.1) will be represented by two pairs of loop categories L1 and L2. One pair, for instance (L1, L1), establishes the link between the signalling

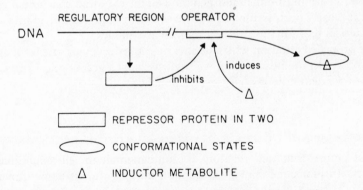

Figure 17.1. Negative inducible regulation of gene expression.

metabolite, i, and protein P. The other pair (L2, L2) indicates the recognition of the operator sequence by protein P.

We require two indexed pairs of L categories, with one empty Li for each i, which will permit the subsequent displacement of the lexical items. We will thus obtain a linear, linguistic representation of regulation loops such as those shown in Figure 17.1. The transformational rules which will actually enable us to represent these loops are the movement rules.

A transformational rule is formed from two descriptions. The first description is a structural description (S.D.) which constitutes the substrate over which the transformation will be applied. The second description is a structural change (S.C.) which constitutes the new structure resulting from the transformation process. For example:

$$\text{S.D.:} \qquad X - \text{Verb} - Y \qquad\qquad (17.9)$$

$$\text{S.C.:} \qquad \text{Verb} - X - Y$$

where X and Y are lexical or syntactic categories. The application of (17.9) over

$$\begin{array}{cccc} (\text{Mary}) & - & (\text{is}) & - & (\text{beautiful}) \\ X & - & \text{Verb} & - & Y \end{array} \qquad (17.10)$$

generates the following interrogative sentence:

$$\begin{array}{cccc} (\text{Is}) & - & (\text{Mary}) & - & (\text{beautiful})? \\ \text{Verb} & - & X & - & Y \end{array} \qquad (17.11)$$

The transformational rules used in describing Figure 17.1 are three rules applied successively, one after the other. We shall write the S.D. of the first and the S.C.s of the rest. We begin with:

$$\begin{array}{ccc} \text{S.D.:} \quad (S, P,) & (Op,) & (S, i) \\ \text{SLL} \quad (, L1, L2) & R(, L1) & SL(, L2) \end{array} \qquad (17.12)$$

Where the L categories and the syntactic categories to which the various parentheses belong are indicated in the lower line. These categories do not change; therefore, we will omit them from this point on. P and i represent the DNA-binding protein and the signalling metabolite, respectively. The first rule generates the following structural change of (17.12):

$$\text{S.C.:} \qquad (S, eP,) (Op, P) (S, i) \qquad (17.13)$$

which represents the binding of P to the operator. Note that each displaced item leaves a trace, e, in its original place. These traces are useful in establishing the description of the principles which govern transformational rules, enabling us to represent a *sequence* of events. The second rule represents the binding of metabolite i to the DNA-binding protein, generating from (17.13):

$$\text{S.C.:} \qquad (\text{S, eP, }) (\text{Op, P-i}) (\text{S, ei}) \qquad (17.14)$$

The conformational change induced by i in P promotes its unbinding from DNA such that expression of the structural region is induced. This liberation is represented by:

$$\text{S.C.:} \qquad (\text{S, eP, P-i}) (\text{Op, eP-i}) (\text{S, ei}) \qquad (17.15)$$

As previously discussed (Collado-Vides, in press), the initial localization of P and i obeys principles which determine the application of transformational rules. Using two principles, we have been able to predict the successive application of the four transformational rules which, in turn, enable us to describe the mechanisms underlying positive and negative, inducible and repressible operon regulation.

From a grammatical standpoint, we proposed the hypothesis stating that the genome determines its own regulation. This hypothesis was incorporated as part of the linguistic analysis method by establishing the determination of transformational rules by the representation derived from phrase structure rules.

PROBLEMS AND PERSPECTIVES

The application of generative grammar concepts to the theoretical study of genetic information organization and regulation has been illustrated. Without going into much detail concerning any of the aspects mentioned, the type of results presented here shows the usefulness of the general justifications underlying the use of the grammatical paradigm.

The basic principles of a theoretical methodology of general applicability in genetics, were briefly exposed. In fact, both phrase-structure rules and transformational rules can, in principle, be applied to any genetic structure at a molecular level.

It is in this sense that we distinguish 'regulability' from 'usefulness', a property which depends on the particular information content of the 'genetic sentences'. The grammatical approach attempts to define rules and principles at a syntactic level; that is, these rules and principles should operate in a manner independent of the specific information contained in genetic 'sentences'.

The elaboration of a grammar specific to a certain genetic structure, such as the *lac* operon, must take into consideration knowledge from other operons, like and unlike, in order to find grammatical rules of general application. In this sense, a grammatical rule is a working hypothesis subject to experimental proof concerning its predictions, and to modification in the face of broader ranging rules.

I have presented an analogy comparing a grammar to a family of equations; in effect, the generative character of grammatical rules provides general descriptions using a finite group of rules for a large amount of data. The existence, *in principle*, of infinite length chains in language, like the family of sequences described in (17.4), shows clearly that generative grammar is a theory dealing with a *capacity* (competence) and not with a corpus, regardless of how large it may be. Generative grammar does not contemplate in the study of English, for example, establishing a grammar capable of deriving all sentences found in the New York Library as a goal, because within the competence criterion, that which is studied is the human capacity for speech, and within this context there is no maximum limit to the length of a sentence.

The selection of the membership criterion based on 'regulability' at a molecular level, determines the infinite nature of the language studied by the grammatical approach. In effect, the chains of such language are products of evolution, and present-day, as well as future species will always be subject to change. Within this context, just as in human language, the number of sentences in genetic language is infinite and, in fact, there is no maximum limit to the length of a genetic chain.

To conclude, I shall mention what I consider to be two important challenges which must be solved by the grammatical paradigm concerning the study of organization and regulation of genetic information:

1. In the first place, it must be shown, through the analysis of a large number of genetic 'sentences', that there are, in fact, general rules describing genomic organization. Such a hypothesis is in contrast to 'Cove's principle' (according to Beckwith, 1987) which states that 'perhaps the most important principle to emerge out of the study of the regulation of gene expression is that general principles do not exist'.

2. Linguistics studies a linear order, such that the methodological challenge of the present approach lies in establishing a 'code' or nomenclature, which may, in turn, establish a linear representation of regulatory events which occur by way of mutually co-ordinated 3-D structures.

The way to begin handling the above mentioned problems, I believe, is by making use of a higher degree of abstraction in the linguistic methodology applied to genetics. In effect, this paper concentrates on illustrating the possible applications of grammatical rules. A future stage of development will probably centre around the search for principles that govern the application of such rules.

Linguistic theory is extremely broad and flexible, more so in comparison to the few methodological applications used in the study of genetics. Consider, for example, the notion of grammatical function as stated by Lyons (1968), in which the Name category, for example, has a Number function (singular, plural) associated with it, the Verb category has a Time function associated with it, etc. These ideas may well be developed in a genetic context.

Generative grammar applied to the study of human language serves as a model, as a source of analogies and as a reference for the study of genetics; however, it by no means determines the specific way in which generative grammar may be applied to genetics. There are no *a priori* conditions concerning the theoretical structure we are seeking. This ample liberty together with the flexibility offered by the generative theory, will hopefully lay the foundations of a methodology for the theoretical study at a molecular level of genetic information that is flexible and free.

REFERENCES

Beckwith, J. (1987) The operon: an historical account. In Neidhardt, F.C. *et al.*, *Escherichia coli and Salmonella Typhymurium. Cellular and Molecular Biology*, vol. 2. American Society for Microbiology.

Chomsky, N. (1957) *Syntactic Structures* (Spanish version) Siglo XXI. México.

—— (1975) *The Logical Structure of Linguistic Theory*. University of Chicago Press.

—— (1980) *Rules and Representations*. Columbia University Press.

Collado-Vides, J. (in press) A transformational-grammar approach to the study of the regulation of gene expression (Journal of Theoretical Biology, 1989) in press.

of gene expression regulation.

Emonds J.E. (1976) *A Transformational Approach to English Syntax. Root, Structure-Preserving and Local Transformations*. New York: Academic Press.

Gould, S.J. (1977) *Ontogeny and Phylogeny*. Cambridge, Mass.: The Belknap Press of Harvard University Press.

Hawley D.K. and McClure W.R. (1983) Compilation and analysis of *Escherichia coli* promoter DNA sequences. *Nucleic Acids Res.11*, 2237–55.

Jackendoff, R. (1977) X̄ Syntax: A study of phrase structure. MIT Press, Cambridge, Mass.

Lightfoot, D. (1982) *The Language lottery. Toward a Biology of Grammars*. MIT Press.

Lyons J. (1968) *Introduction to Theoretical Linguistics*. Cambridge Unievrsity Press.

van Riemsdijk, H.C. and Williams E. (1986) *Introduction to the Theory of Grammar*. MIT Press.

AUTHOR INDEX

SUBJECT INDEX

acquisitive evolution, 60–2
adaptation
  of complex systems, 124–33
  of immune networks, 12–23
adaptive landscape, 68–77, 81–7
amino acids, categorizations of, 206–7
arrow of time, 109–110

bifurcations
  and epigenetic landscape, 16–30, 97–8
  in reaction-diffusion systems, 33–40

CFSs, 112, 119–22
calcium, and phase transitions of
  phospholipids, 153–4, 158–9
canalization, 3, 4
catastrophe theory, 5, 18–20
  butterfly catastrophe, 24–8
  cusp catastrophe, 20–4
cell-cell interactions, discrete aspects, 181–7
chaotic systems, 103, 106
Cholesterol and phase transitions, 153
Classifier Systems, 112, 119–22
coherent excitations, 162–76
colour patterns
  butterflies, 184–5
  snakes, 184
complex systems
  and adaptation, 124–32
  physics of, 101–111
  spontaneous order in, 67–80
continuum mechanics and physiology, 191–7
contraction of muscles, 163–4, 169–70
creod, 3–4

DNA and language theory, 207, 208
demand theory of gene regulation, 48–63
development an evolution, 89–100
differentiation and demand theory, 56–60
Drosophila
  epigenetics, 5
  homeotic bands, 185–7
dysrhythmic cerebral state, 195

E. coli *see* Escherichia coli
epigenetic landscape, 4–5
  bifurcations and, 16–30, 97–8
epigenetics, 2–3
Escherichia coli
  gene regulation, 43, 46, 49–50, 52–4, 57–8,
    60–1
  phase transition, 153
evocation, 2
evolution and development, 89–100
evolution, acquisitive, 60–2
excitability of membranes, 157–8
excitations, coherent, 162–76

field models, 134–45
fields, gradient, classification of, 10–12
fractal structures, 106–7

gastrulation, 92–6
gene dyamics, discrete aspects, 178–81
gene regulation, 42–66, 77–87
  see also regulability
generative grammar *see* grammar
generic equations, classification of, 8–15
gradient fields, classification of, 10–12
grammar
  definitions of, 204–5, 213
  generative, 204–5, 211–224; defined, 213
  phrase-structure, 205–6, 218–20

homeorhesis, 6, 17
homeotic bands, in Drosophila, 185–7

immune networks, adaptiveness in, 112–23
individuation, 2
interactions
  cell-cell, 181–7
  long range, 187–90
  short range, 178–81

kindling
  effect, 193–6